Alexander A. Friedn

The World as Space and Time

Translated by Svetla Kirilova-Petkova
and Vesselin Petkov

Edited by Vesselin Petkov

MINKOWSKI
Institute Press

Alexander Friedmann
29 June 1888 – 16 September 1925

Cover: Based on Fig. 20 of the book.

ISBN: 978-1-927763-19-3 (softcover)
ISBN: 978-1-927763-20-9 (ebook)

Minkowski Institute Press
Montreal, Quebec, Canada
http://minkowskiinstitute.org/mip/

For information on all Minkowski Institute Press publications visit our
website at http://minkowskiinstitute.org/mip/books/

Preface

The book of the great Russian physicist Alexander A. Friedmann *Мир как пространство и время* – *The World as Space and Time* – was first published in the original Russian in 1923 by the publisher *Academia* in what was then Petrograd (later Leningrad, and now back to St. Petersburg). Although the book had been initially planned as an article for the philosophical journal *Thought*, it is one of the first introductions to the spacetime physics of the theory of relativity for a wider audience. Friedmann had succeeded in both making the book accessible to non-experts and providing rigorous explanations.

This is the first English translation of Friedmann's book and I hope it corrects an unfortunate omission to publish in English the non-technical book by the physicist who first showed in 1922 that Einstein's equations have solutions that describe a non-stationary Universe (later the experimental evidence did confirm that the Universe is expanding).

In order to reduce the Editor's notes to a minimum I decided not to comment on most technical issues in the book (including Friedmann's terminology) since such comments would be of little help to non-experts, whereas experts do not need them.

Svetla and I would like to thank Sergey Andronenko (St. Petersburg, Russia) for the translation of several passages in the book, mostly Russian poems.

Montreal, 23 March 2014   *Vesselin Petkov*

Editor's Introduction[1]

If the talented Russian physicist Alexander Alexandrovich Friedmann had to be introduced with a single sentence, the most appropriate sentence would be the title of his biography translated from the Russian: *Alexander A. Friedmann: The Man who Made the Universe Expand.*[2]

Indeed, he was the first to realize in 1922 that Einstein's equations have solutions which describe not only a stationary Universe as Einstein initially believed, but also a non-stationary world. Friedmann won the debate with Einstein over the admissibility of such solutions, but his life was too short and he could not see the triumph of his views when the experimental evidence fully supported his predictions and demonstrated that the Universe was expanding.

Friedmann was born on June 29, 1888 in St. Petersburg.[3] His parents had very little to do with science. His father was a ballet dancer, musician and composer, and his mother (the daughter of a famous musics professor in the St. Petersburg Conservatory, composer and conductor) gave piano lessons. Friedmann admitted he did not inherit any musical talent from his parents.

From 1897 to 1906 Friedmann attended the Second St. Petersburg Gymnasium where he impressed his mathematics teachers with his

---

[1] As this Introduction summarizes Friedmann's scientific and personal life it is unavoidably essentially the same as the Introduction to Alexander A. Friedmann, *Papers On Curved Spaces and Cosmology* (Minkowski Institute Press, Montreal 2014); here I added what Friedmann learned about the sad consequences of his research on the theory of precision bombing.

[2] E. A. Tropp, V. Ya. Frenkel, A. D. Chernin, *Alexander A. Friedmann: The Man who Made the Universe Expand.* Translated by A. Dron and M. Burov (Cambridge University Press, Cambridge 1993).

[3] This summary of Friedmann's biography is based on P. Ya. Polubarinova-Kochina, In Memory of A. A. Fridman (On the seventy-fifth anniversary of his birth), *Usp. Fiz. Nauk* **80** (1963) pp. 345-352, V. Ya. Frenkel, Alexander Alexandrovich Friedmann (Biographical sketch) *Usp. Fiz. Nauk* **155** (1988) pp. 481-516, and E. A. Tropp, V. Ya. Frenkel, A. D. Chernin, *Alexander A. Friedmann: The Man who Made the Universe Expand.* Translated by A. Dron and M. Burov (Cambridge University Press, Cambridge 1993).

mathematical talent. In 1905, while still at the school, Friedmann and his friend Ya. D. Tamarkin wrote their first scientific paper on the Bernoulli numbers, which was published in 1906.

Not only was Friedmann scientifically active, but he was also involved in the political life of that time. In 1905-1906 he took part in the students' movement of the St. Petersburg secondary schools and was a member of the Central Committee of the student organization.

From 1906 to 1910 Friedmann was a student at the Faculty of Physics and Mathematics of St. Petersburg University. After graduating from the University in 1909, he was invited by two of his professors – V. A. Steklov and D. K. Bobylev – to pursue advanced studies with the possibility to be retained at the University as a professor in the Department of Pure and Applied Mathematics.

Friedmann (first row, second from left) with colleagues and friends. His wife Ekaterina is first (from left) in the second row.

In 1911 Friedmann married Ekaterina Petrovna Dorofeyeva which got by surprise his friends, colleagues and professors. On 13 July 1911 his friend Tamarkin found it nevessary to inform their professor Steklov:[4]

> The marriage of Alexander Alexandrovich was as unexpected to me as it was to you. His wife is quite a good-looking woman, although slightly older than he is. So far, I can say that the marriage has had only a positive effect on Alexander; it has reduced his habitual nervousness, made him calmer, and in no way hampered our studies, which have continued without interruption five times a week.

[4]E. A. Tropp, V. Ya. Frenkel, A. D. Chernin, *Alexander A. Friedmann: The Man who Made the Universe Expand.* Translated by A. Dron and M. Burov (Cambridge University Press, Cambridge 1993), p. 55.

Friedmann and his wife often visited Steklov's home which allowed Steklov to note that Ekaterina was a very quiet woman who had a considerable and beneficial effect on her husband. Later, in 1925, "Steklov applied for a pension for Ekaterina Friedmann and highly commended the assistance that she had given to Alexander Friedmann, working on translations of his articles, reading proofs, etc."[5]

In 1914 Friedmann is sent to Leipzig in Germany to do scientific research work under V. Bjerknes.

From 1914 to 1916 Friedmann took part in World War I. He volunteered to join the Army and served in aviation units of the northern and southern fronts.

Even during the war Friedmann did not stop doing research – as a pilot he worked on the theory of precision bombing, compiling the appropriate tables, the use of which had risen dramatically the probability for the bombs to hit the intended target. In a letter to Steklov (dated 28 February 1915) Friedmann discussed his work on the theory of precision bombing:[6]

> I have recently had a chance to verify my ideas during a flight over Przemyśl; the bombs turned out to be falling almost the way the theory predicts. To have conclusive proof of the theory I'm going to fly again in a few days. The bombs I drop (5 lb, 25 lb and 1 pood [40 lb] in weight) belong to the class in which α is very small, so I've been verifying the expansion of the solutions in terms of the parameter α.

As it often happens in times of war such an episode had a sad and curious continuation – later Friedmann learned that one of his bombs destroyed the house of one of his meteorology colleagues who was a German pilot at the time:[7]

> Przemyśl ... was a strongly fortified Austrian fortress in Western Galicia. It was blockaded by Russian troops in late September 1914, seized by them on March 22, 1915, and abandoned two months later. So Friedmann's flights over Przemyśl were made during its six-month blockade.

---

[5] *Ibid.*

[6] *Ibid*, p. 72.

[7] *Ibid*, pp. 72-73.

It so happened that H. Ficker, a future professor of meteorology (whom Friedmann was destined to meet in peacetime in 1923) was in 1915 at Przemyśl, then occupied by Austrian troops. In Friedmann's obituary, V. A. Steklov relates that in 1925, on learning from Steklov about Friedmann's death, Ficker wrote back saying that the only bomb he saw hit the target in Przemyśl was dropped from the airplane Friedmann was in.

A natural question arises: how could Ficker know this? The answer is given by the quotation from Ficker's extremely sympathetic interview which he gave to a correspondent of the Leningrad *Krasnaya Gazeta* (Red Newspaper): "I met with Friedmann three times in Berlin. One day we had a talk and it turned out that not long before the fall of Przemyśl Professor Friedmann, being a Russian military pilot, dropped a very powerful bomb over my home there. When the bomb fell, being a German pilot, I was in my superior's office to receive my orders. I remember well that the bomb which Friedmann dropped was the only Russian bomb which hit the target at Przemyśl. When I first met Friedmann in Berlin we found out the exact time and place of our unusual and unfriendly acquaintance on the battlefield."

After the war Friedman held a number of positions, including Professor and Chair of the Department of Mechanics at the Perm State University (1918-20), Professor in the Faculty of Physics and Mathematics of the Petrograd Polytechnical Institute (1920-25) and from February 1925 – Director of the Main Physical (later Geophysical) Observatory.

Friedmann in his office in the Main Physical Observatory.

In 1923 had enormously difficult time (in his own words) when he divorced his wife Ekaterina and married Natalia Yevgenievna Malinina (1893-1981) – a physicist at the Main Physical Observatory who worked at the time on Earth's magnetism and, particularly, the Kursk magnetic anomaly.

In his scientific career Friedmann worked mostly in the fields of hydrodynamics and metereology. In the last several years of his life he directed his intellectual power toward the theory of relativity and its implications for cosmology. As indicated in Friedmann's Biography he and his colleague V. K Frederiks had been intensely working on the theory of relativity[8]:

> From the 1920s there was a regular seminar at the Physical Institute of Petrograd University, where Friedmann and Frederiks presented papers on general relativity.

On 15 April 1922 Friedmann completed his manuscript "On the Geometry of Curved Spaces" and sent it to Ehrenfest. A bit later (on 29 May 1922) he completed his paper "On the curvature of space", which appeared in *Zeitschrift für Physik* the same year.[9]

In 1923 Friedmann published the book *The World as Space and Time* on the emerging (after the works of Einstein and Minkowski) spacetime view of

Friedmann with his second wife Natalia Malinina (left)

the world, which was written for a wider audience. In the book Friedmann demonstrated amazing understanding of Minkowski's revolutionary ideas outlined in his talk "Space and Time".[10]

Both Minkowski and Friedmann left this world prematurely and were not able to contribute to our understanding of the physical world what they could have done. Had they lived longer physics would certainly be different.

In November 1923 Friedman completed his paper "On the Possibility of a World with a Constant Negative Curvature of Space", which appeared in *Zeitschrift für Physik* the following year.[11]

In 1924 Friedmann and Frederiks began to write a fundamental monograph on the theory of relativity[12]:

The cover of Friedmann's book *The World as Space and Time.*

---

[8] E. A. Tropp, V. Ya. Frenkel, A. D. Chernin, *Alexander A. Friedmann: The Man who Made the Universe Expand.* Translated by A. Dron and M. Burov (Cambridge University Press, Cambridge 1993) p. 115.

[9] A. Friedman, Über die Krümmung des Raumes, *Zeitschrift für Physik* **10** Nr. 1, 1922, S. 377-386.

[10] In: H. Minkowski, Space and Time. New translation in: H. Minkowski, *Space and Time: Minkowski's Papers on Relativity* (Minkowski Institute Press, Montreal 2012).

[11] A. Friedmann, Über die Möglichkeit einer Welt mit konstanter negativer Krümmung des Raumes, *Zeitschrift für Physik* **21** Nr. 1, 1924, S. 326-332.

[12] E. A. Tropp, V. Ya. Frenkel, A. D. Chernin, *Alexander A. Friedmann: The*

They set themselves the task of presenting the theory with
adequate rigor from the logical point of view, assuming
the reader's background in mathematics and theoretical
physics did not exceed the level of knowledge given by
Russian universities and higher technical educational in-
stitutions. It was originally intended to publish the whole
book at once, but technical obstacles made the authors
divide the book into five parts and prepare these parts
as separate issues. The first issue of the book expounded
the fundamentals of tensor calculus. The second issue was
to be devoted to the fundamentals of multi-dimensional
geometry, the third to electrodynamics, and, finally, the
fourth and fifth to the fundamentals of special and general
relativity.

Unfortunately, only the first volume of the
monograph was published:[13] V. K. Frederiks
and A. A. Friedmann, *Foundations of the Theory of Relativity, Volume 1: Tensor Calculus*
(Academia, Leningrad 1924).

The first reaction to Friedmann's work
came from Einstein himself and it was negative. In a brief note[14] which was published in
Zeitschrift für Physik in 1922 he wrote:[15]

The work cited contains a result
concerning a non-stationary world
which seems suspect to me. Indeed,
those solutions do not appear com-
patible with the field equations (A).
From the field equations in follows

The first and only
published volume of
*Foundations of the
Theory of Relativity* by
Friedmann and Frederiks.

necessarily that the divergence of the matter tensor T_{ik}
vanishes. This along with the anzatzes (C) and (D) leads

Man who Made the Universe Expand. Translated by A. Dron and M. Burov (Cambridge University Press, Cambridge 1993) p. 117.

[13]The first English translation of this book will appear in 2014. See V. K.
Frederiks and A. A. Friedmann, *Foundations of the Theory of Relativity, Volume
1: Tensor Calculus* (Minkowski Institute Press, Montreal 2014) at http://www.
minkowskiinstitute.org/mip/books/friedmann3.html

[14]A. Einstein, Bemerkung zu der Arbeit von A. Friedmann Über die Krümmung
des Raumes, *Zeitschrift für Physik* **11** (1922), p. 326.

[15]Quoted from: J. Bernstein and G. Feinberg (eds.), *Cosmological Constants:
Papers in Modern Cosmology* (Columbia University Press, New York 1986) p. 66.

to the condition

$$\frac{\partial \rho}{\partial x_4} = 0$$

which together with (8) implies that the world-radius R is constant in time. The significance of the work therefore is to demonstrate this constancy.

On 6 December 1922 Friedmann sent a detailed letter to Einstein. Here is part of it:[16]

Dear Professor,

From the letter of a friend of mine who is now abroad I had the honor to learn that you had submitted a short note to be printed in the 11th volume of the *Zeitschrift für Physik*, where it is stated that if one accepts the assumptions made in my article "On the curvature of space," it will follow from the world equations derived by you that the radius of curvature of the world is a quantity independent of time.

. . .

Should you find the calculations presented in my letter correct, please be so kind as to inform the editors of the *Zeitschrift für Physik* about it; perhaps in this case you will publish a correction to your statement or provide an opportunity for a portion of this letter to be printed.

Friedmann's "a friend of mine" was Yu. A. Krutkov. In May 1923 Krutkov met Einstein at Ehrenfest's home in Leiden and discussed Friedmann's letter with him. In particular, Krutkov drew Einstein's attention to that part of the letter where Friedmann showed by direct calculations that the necessary condition for the disappearance of the divergence of the matter tensor, which was pointed out by Einstein in his note, by no means implies that the radius of the curvature of the world is constant in time. Both Friedmann's letter and Krutkov's explanation convinced Einstein that the results of Friedmann's paper were correct and on 21 May 1923 he sent a second note[17] concerning Friedmann's paper to *Zeitschrift für Physik*:[18]

[16]E. A. Tropp, V. Ya. Frenkel, A. D. Chernin, *Alexander A. Friedmann: The Man who Made the Universe Expand*. Translated by A. Dron and M. Burov (Cambridge University Press, Cambridge 1993) p. 170.

[17]A. Einstein, Notiz zu der Arbeit von A. Friedmann "Über die Krümmung des Raumes," *Zeitschrift für Physik* **16** (1923), p. 228.

[18]Quoted from: E. A. Tropp, V. Ya. Frenkel, A. D. Chernin, *Alexander A. Friedmann: The Man who Made the Universe Expand*. Translated by A. Dron and M. Burov (Cambridge University Press, Cambridge 1993) p. 172.

In my previous note I criticized the above-mentioned work. However, my criticism, as I became convinced by Friedmann's letter communicated to me by Mr. Krutkov, was based on an error in calculation. I consider that Mr. Friedmann's results are correct and shed new light. It has turned out that the field equations allow not only static but also dynamic (i.e. variable with respect to time) centro-symmetrical solutions for the space structure.

Despite that Einstein admitted that his equations have solutions for non-stationary worlds, he did not believe that such solutions reflect something real. During the fifth Solvay meeting at Brussels in 1927 Einstein told Lemaître that in his opinion non-stationary world models are "simply disgusting."[19]

Friedmann in 1922 or 1923.

After the successful outcome of the debate with Einstein, the future looked bright and promising for Friedmann. Unfortunately, he had two more years to live. One of the last events he took part was in July 1925 when he participated in a balloon flight which reaching the record elevation of 7,400 m. After that Fiedmann and his wife went on vacation to Crimea. After he returned to Leningrad he became ill. It turned out that when returning from Crimea in August 1925 Friedmann bought and ate pears without washing them and contracted a deadly disease – typhus:[20]

Weakened by the deprivations of the war years and his exhausting work, he failed to throw off the disease, and on September 16, Alexander Alexandrovich Friedmann died.

Friedmann did not live to see the birth of his son:[21]

---

[19]T. Jung, Three cosmological dogmas – Einsteins influence on early relativistic cosmology, *Astron. Nachr.* / AN 326 (2005), No. 7, p. 1

[20]E. A. Tropp, V. Ya. Frenkel, A. D. Chernin, *Alexander A. Friedmann: The Man who Made the Universe Expand.* Translated by A. Dron and M. Burov (Cambridge University Press, Cambridge 1993) p. 211.

[21]*Ibid,* p. 209

Friedmann's son, the third Alexander Alexandrovich in their family, used to say to his mother when he was a teenager: "I want to be an ordinary person." He entered university during the war, served in the Army as a driver, and worked as a driver in Leningrad after the war, until his death in 1983. He had no children.

Let me finish with a final quote from Fridmann's Biography:[22]

> Ekaterina Petrovna Dorofeyeva-Friedmann writes in her memoirs: "Always ready to learn from everybody who knew more than he did, he realized that in his work he was blazing new trials, difficult and unexplored by anyone, and he liked to quote these words of Dante's: 'L'acqua ch'io prendo giammai non si corse' (*Paradiso*, II, 7). 'The sea I sail was never crossed before'."

> In the pages of his research, in the reminiscences of his contemporaries, Friedmann is seen as a profound, independent-minded, and daring thinker who destroys scientific prejudices, myths and dogmas; his intellect sees what others do not see, and will not see what others believe to be obvious but for which there are no grounds in reality. He rejects the centuries-old tradition which chose, prior to any experience, to consider the Universe eternal and eternally immutable. He accomplishes a genuine revolution in science. As Copernicus made the Earth go round the Sun, so Friedmann made the Universe expand.

---

[22] *Ibid*, p. 175

CONTENTS

xiv

INTRODUCTION

Once, when night covered the heavens with its coat, fa-
mous French philosopher Descartes, seating near his home
stairs and looking at gloom horizon with great attention,
– a passerby approached him with a question: 'Say, sage,
how many stars are on this sky?" – "Rascal! *– the sage*
answered, – nobody can embrace the boundless!" *These*
emphatic words produced the desired effect on the passerby.

"Historical materials of F.K. Prutkov (grandfather)"[23]

1. In the above exchange with Descartes the passerby "got wiser"
and calmed down. But in reality in the human history the pursuit to
"count the stars," or, in other words, to construct a world picture has
never ceased regardless of the level of people's knowledge there have
always existed individuals, ordinary people and great thinkers alike,
trying to create the picture of our Universe on the basis of always
insufficient data.

In the twentieth century humans tried again, based on the knowledge gathered by the natural sciences by that time, to create a general
picture of the world. It is true that this picture was extremely simple and resembling the real world as much as the dim reflection in
the mirror of a schematic drawing of a famous building resembles the
building itself. That attempt to "count the stars" and create a general
picture of the world have the not very adequate name – the *principle
of relativity.*

2. The world, whose schematic picture is created by the principle
of relativity, is the world of the natural scientists, is a class of only such
objects which can be described or evaluated with numbers. For this
reason, we could say that this world is infinitely narrower and smaller
than the world-universe of philosophers. And if it so, of course, the
importance of the relativity principle for philosophy should not be

---

[23]EDITOR'S NOTE: Translated by Sergey Andronenko, St. Petersburg, Russia.

overestimated. Probably after all, the importance of the relativity principle for philosophy is not greater than the importance for it of the cosmogonic hypotheses of modern astronomy. But at the same time we should not jump to another extreme conclusion and deny completely the philosophical significance of the world picture, given by the relativity principle. The great and deep thoughts, characterizing the general concepts and ideas of the relativity principle, concerning such objects as space and time (it is true measurable) should make no doubt certain impression if not even influence on the development of the ideas of the contemporary philosophers, who often stay much higher above "the measurable" Universe of the natural scientists.

3. The discussed topic above makes me think, that gaining some *real* knowledge of the principle of relativity on the pages of a philosophical journal[24] could be helpful and useful. Speaking of real knowledge I am against some attempts and explanations, which try to provide a superficial popularization of the relativity principle, which could not be popularized without extreme care. Such popularizations are usually achieved at the price of a complete obstruction of the relativity principle and, even worse, at the price of abuses of its ideas and scientific meaning. From this, it is clear, I hope, that this work does not in any case pretend to be popular and requires some knowledge in advanced mathematics in order to be understood.

4. I divided this work into three chapters. The first chapter is devoted to the theory of the general properties of space. To simplify and to make it clear I discuss a space of three, and even often, two dimensions, but all conclusions can be extrapolated, by means of simple although sometimes not rigorous analogies, to a space of any number of dimensions, including a four-dimensional space. I begin this chapter with an analysis of the question of dimensions, which is extremely important for the correct understanding of the many conceptions of the theory of relativity. Discussing further the basic properties of the geometrical space (metrics, direction, angle, parallel transport, straight line, curvature, etc.), I will explain, in detail, the geometrical interpretation of space with the help of the physical space. Pointing out the arbitrariness of this interpretation I will also show what the significance of the experimental study of the properties of space is. It should be noted that this issue, of course, has no meaning if the geometrical interpretation of space is not established.

The second chapter deals with one of the most difficult questions of theory of relativity, namely of time and the impossibility of the inde-

---

[24]This work had been written for the journal *Thought*.

pendent existence of time and space separately, and of their unification into a four-dimensional physical world, which can, in a certain sense, interpret the geometrical world. It is self-evident, that in this chapter we discuss not the time of philosophers, but the modest "measurable" time of the natural scientist. A special attention is devoted to the detailed clarification of the complete arbitrariness of the definition of this "measurable" time.

The world, i.e. the unification of space and time, is discussed in great detail in the second chapter. A particular attention is devoted to the *postulate of invariance* which gives the correct idea when the physical laws of our world are determined. The postulate of realness and the connected with it causality are also well explored. No less attention is given to the difference between the physical and the geometrical world and the arbitrariness of the interpretation of one of the worlds through the other. Note, that the historic unification of space and time into a four-dimensional world became possible with the help of the so called special principle of relativity. Logically, that principle can be completely left out without any harm to our explanation. And I will do it taking into account that this principle of relativity has been often discussed in detail in the general and the special scientific literature alike.

The third chapter will describe methods, with the help of which the relativity principle tries to build the picture of the world. I will start this chapter by exposing the foundations of the old and new mechanics, mainly by establishing the law of inertia, having place in the new general mechanics. After that I will discuss several hypotheses which helped Einstein connect the force of gravity with the geometry of the world. The experimental confirmation of these hypotheses demonstrate their correctness at least generally.

The link of gravity with the properties of our world is one of the greatest ideas of Einstein, although the origin of this idea comes from earlier times, namely from the time when the famous work of Riemann appeared. Starting with Riemann's ideas, using Hilbert's ideas and generalizing and expanding Einstein's ideas, the German mathematician Weyl suggested that between the properties of matter, filling the world, and the properties of the world there exists a close connection, making it possibly, through experimenting with matter, to determine in detail the properties of the world. I think, Weyl's ideas should be given a deserved place in the third chapter.

The end of the third chapter will be devoted to the general structure of our (self-evidently, material) Universe. Regardless of how shaky our considerations concerning this field are, the importance of this task

4

requires to deal with it with great interest. I think this question (of the construction of our Universe) should be elucidate in great detail, because due to the fashion of the relativity principle and the huge amount of popular books and lectures dealing with this principle, not only in the "society," but in more professional circles, completely distorted information about finiteness, curvature and other properties of space, as if established by the relativity principle, has been circulating.

5. In order to help the reader avoid a wrongful impression about this work, the author should warn, that he is not in any case a philosopher and explains the relativity principle from a purely mathematical point of view. It is very possible that this is the only point of view which can explain (more or less) clearly the foundations of the relativity principle.

It is useless to recommend any literature on different aspects of the relativity principle, because the popular literature does not explain anything, and the special literature requires the necessary mathematical knowledge. Nevertheless, I will recommend several special books and articles, devoted to the relativity principle.

1. H. Weyl, *Raum – Zeit – Materie*. Vierte Auflage. Berlin, 1921.

2. M. von Laue, *Die relativitätstheorie*. Zweiter Band. Braunschweig, 1921

3. A. S. Eddington, *Espace, Temps et Cravitation*. Paris, 1921.

4. D. Hilbert, *Die Grundlagen der Physik*. Göttinger Nachrichten. 1916.

SPACE

God created the whole existing world accordingly to measure and quantity.

"Book of Solomon's wisdom"[25]

§1 Measuring Quantities

1. Considering different *properties* of the material world, we can group these properties in special *classes*, assigning to a given class properties with some common feature. The properties assigned to a class form a *manifold*, which may possess different cardinality, that is, may be in most cases put in one-to-one correspondence either with the manifold of integers, or with the manifold with real numbers, or with the manifold of pairs, triples, etc. of real numbers. Each property of a class will be assigned one and only one number (or one and only one pair of numbers, or one and only one triple of numbers, etc.) of the manifold of numbers, and vice versa. The setting of rules, by which we can find the number (or a pair, or a triple of numbers) which corresponds to a given property of a class, and find the property of a class, which corresponds to a given number (or a given pair, or triple, etc., of numbers), we will call, for the sake of brevity, *the arithmetization of a given class.*

Let us give a pair of simple examples. People possess the property of gender; assign the number 0 to the property of a person to be a woman[26], and the number 1 to the property of a person to be

---

[25] EDITOR'S NOTE: Translated by Sergey Andronenko, St. Petersburg, Russia.

[26] EDITOR'S NOTE: After the appearance of the first Russian edition of *The World as Space and Time* in 1923, a colleague of Friedmann confronted him and expressed her indignation that the property of a human being to be a woman had been assigned the number 0. Friedmann, who had been well-known for his fair treatment of everyone regardless of gender or nationality, promised to reverse the assigned numbers 0 and 1 in the second edition of the book. Unfortunately, he died

6

a man; the setting of this rule will give us the arithmetization of this class of properties. All material bodies possess the property of having a volume; by defining the volume using conventional means in cubic meters, we find the rule of assigning numbers to the volumes of objects, otherwise speaking, we obtain the arithmetization of the class of volumes.

2. The arithmetization of a given class of properties can be done always completely arbitrary. Let the properties of a given class be such that they can be characterized by the notions "more" and "less" (which are, of course, defined); we will call the properties of such a class *intensities*. If the arithmetization of a given class is done in such a way so that a greater intensity corresponds to a larger number, such arithmetization is called an *estimatation*.[27] The degree of knowledge acquired by students is intensity (assuming that we are able to determine a greater or lesser degree of knowledge) – so a five- or twelve-point grading system is an arithmetization of the class of degree of student knowledge, which we call estimation (grade).

It is self-evident that each class of properties can become a class of intensities, as soon as we define, somehow, the notions "more" and "less" applicable to the properties of a given class. Most properties, which we study, refer directly to the category of intensity, since the very definition of a given class of properties already contains the notions "more" and "less" characterizing the properties of this class. Sometimes, however, the properties are expressed in such a way, that their representation as a class of intensities requires additional definitions. An example is the property of a mono-chromatic light ray to have a color; at first sight, it seems meaningless to assume that the color has intensity, but introducing an additional definition we can consider that a color is greater if it has a longer wavelength; as soon as this definition is introduced – the color will become intensity, and it will be easy to arithmetize this class of properties, so the arithmetization becomes estimation. Therefore, the representation of a class of properties as a class of intensities depends completely on our definitions, and, therefore, on our will. Of course, I will not touch on the question of the reasonableness of the applicability of the notion intensity to one or another property.

---

in 1925 and did not see the second edition of his book. See: P. Ya. Polubarinova-Kochina, In Memory of A. A. Fridman (On the seventy-fifth anniversary of his birth), *Usp. Fiz. Nauk* **80** (1963) pp. 345-352, p. 347.

[27] EDITOR'S NOTE: The Russian word is "оценка" whose general meaning is "estimation;" in the next sentence, however, it has a narrower (specific) meaning and should be translated as "grade."

3. Among the intensities,there is a special group, called dimension, which possesses characteristic properties making it possible to have a particular estimation (arithmetization). Let $A_1, A_2, A_3, \ldots$ be the intensities of a given class A; as we deal with intensities, we have a definition of the meaning of the assertion: A_1 is smaller than A_2. Let us assume that for the intensities of the class A we, one way or another, define the notion of *three equidistant intensities*, where under three equidistant intensities we will understand such intensities A_1, A_2, A_3, so that A_1 is as smaller than A_2, as A_2 is smaller than A_3. Let us call the intensities of the class A, to which the notion of equidistant intensities (defined by us in a certain way) is applicable, *measurable intensities*.

It is self-evident, that by introducing a suitable definition we can transform any class of intensities into a class of measurable intensities. Like the case of representing properties as intensities, the present case also contains two possibilities: either, we will define the notion of equidistant intensities in the very identification of some class of intensities, and then from the very beginning we will deal with measurable intensities; or, we will not define the notion of equidistant intensities in the identification of the class of intensities, and therefore such a class of intensities can be transformed into a class of measurable intensities only after additionally introducing the notion of equidistant intensities. As an example of intensity, from its very identification as a measurable intensity, can serve the volume of a material body or the length of a segment of a straight line. In fact, by introducing the notion of a volume or of a length of a segment of a straight line, from the very beginning we define what we should mean by a larger volume or a longer length, and also what we should mean by three equidistant lengths of three segments. Conversely, the intensity of the degree of student knowledge, from its very definition, is not measurable, and up to now the notion of equidistant degrees of student knowledge has not been defined.

4. Let arithmetize (estimate) the class of measurable intensities, but this estimation will be made in such a way, that successive differences between numbers, representing any three equidistant intensities, would be equal among themselves; in other words, if a_1, a_2, a_3 are numbers, representing in the estimation three equidistant and increasing intensities A_1, A_2, A_3 , then $a_2 - a_1 = a_3 - a_2$.

We will call such a kind of estimation of a class of measurable intensities *dimension*. We saw that the arithmetization of a given class of properties can be done completely arbitrarily; in the estimation of a given class of intensities the arbitrariness is to a certain extent smaller, but nevertheless extremely significant. Let us see what arbitrariness can be involved in the measurement of a given class of measurable intensities. Let a given class A of measurable intensities be measured by employing two different methods: in the first method the numbers $a_1, a_2, a_3, \ldots$ correspond to the intensities $A_1, A_2, A_3, \ldots$, in the second method – the numbers $\bar{a}_1, \bar{a}_2, \bar{a}_3, \ldots$ It follows from the property of arithmetization that $\bar{a} = f(a)$, where f is a certain single-valued function of a. If A_1, A_2, A_3 are three equidistant quantities, then by the basic property, $a_3 - a_2 = a_2 - a_1$ and also $\bar{a}_3 - \bar{a}_2 = \bar{a}_2 - \bar{a}_1$. Therefore, if

$$a_2 = \frac{a_3 + a_1}{2}$$

then

$$\bar{a}_2 = \frac{\bar{a}_3 + \bar{a}_1}{2}$$

But since $\bar{a} = f(a)$ we have

$$f(a_2) = \frac{f(a_3) + f(a_1)}{2}$$

or

$$f\left(\frac{a_3 + a_1}{2}\right) = \frac{f(a_3) + f(a_1)}{2}$$

for all a_1, a_3. Replacing a_1 with x, a_3 – with y, we find that for all x and y we will have the following functional relation:

$$f\left(\frac{x + y}{2}\right) = \frac{f(x) + f(y)}{2}. \tag{1}$$

We can prove that functional equation (1) can be presented in the form:

$$f(a) = \mu a + h,$$

where μ and h are constant numbers.

Obviously, the above expression for $f(a)$ satisfies equation (1)

$$\mu \frac{x + y}{2} + h = \frac{1}{2}(\mu x + h + \mu y + h) = \frac{1}{2}\mu(x + y) + h.$$

In such a way, we found the general form of $f(a)$, making it possible to go from one method of measurement to another:

$$\bar{a} = f(a) = \mu a + h. \tag{2}$$

This general formula of the transition from one method of measurement to another contains two constants. The constant h reflects the arbitrariness of the *initial value*; in other words, the arbitrariness of that intensity, which corresponds to the zero value of the number representing it.

The constant μ reflects the arbitrariness of *the units of measurement*; in other words, the arbitrariness of those two intensities, the difference of whose representing numbers is equal to 1. It is easy to see that by specifying the initial value and the unit of measurement, we can define completely and unambiguously both constants μ and h in equation (2); in other words, we can define unambiguously the measurement of the intensities of a given class. In fact, let the intensity A_0 of the class A be the initial value (i.e. the measurement of that intensity should give zero), and let the intensities A_1 and A_2 define the unit of measurement (i.e. let the measurement of the difference of the numbers representing A_1 and A_2 be equal to 1). Let us perform some kind of measurement of class A. Assume that this measurement of the intensities A_0, A_1, A_2 produces the numbers a_0, a_1, a_2, then the sought measurement gives the numbers $\bar{a}$, connected by the relation (2) to the obtained, from the introduced measurement, numbers a. It remains to choose μ and h in such a way that $\bar{a}_0 = 0$, $\bar{a}_2 - \bar{a}_1 = 1$; in other words, that the following equations be satisfied:

$$\mu\, a_0 + h = 0, \quad \mu\,(a_2 - a_1) = 1,$$

from where

$$\mu = \frac{1}{a_2 - a_1}, \quad h = -\frac{a_0}{a_2 - a_1}.$$

5. In the usual practice of measurement very often we encounter the arbitrariness of units of measurement and much less frequently the arbitrariness of the initial value. This is a result of the practical convenience to choose as the initial value such an intensity of a given class, which possesses some special properties. For example, when measuring the length of a segment, we always take for the initial value the length of a line segment between coincident points (the length of a point), whereas for the unit of measurement we take various differences of lengths (arshins, feet, meters, etc.).

There exists, however, an intensity in which practice has legitimized a large variety of initial values – this intensity is temperature. In fact, for the measurement of temperature in degrees of Celsius and Rømer the temperature of the melting of ice is taken as the initial value, whereas for the measurement of temperature in degrees of Fahrenheit we assign the number 32 to the temperature of the melting ice, and for the measurement of temperature in degrees of the absolute scale – the number 273. It is obvious, that the unit of measurement of the temperature also changes: in the case of degrees of Celsius, the difference of the numbers, representing the temperatures of boiling water and melting ice, is taken to be 100, and in the case of degrees of Rømer this number is 80.

6. I have discussed the question of measurement in such a detail for two reasons. On the one hand, later on we will have to confront repeatedly the process of simple arithmetization with the process of measurement, and, on the other hand, in many courses in mechanics and physics, which have become classic, there exists significant reticence on the question of measurement, whose essence is so simple; my task was namely to establish greater certainty in this matter and at the same time to show how great arbitrariness is in process of measurement.[28]

So, every class of properties can be arithmetized; if these properties are represented (through our definition) by intensities, then not only can we arithmetize them, but also express them by numbers. Finally, if the intensities of a given class are represented (again through our definition) by measurable intensities, then not only can we express them by numbers, but also measure them; the measurement involves certain arbitrariness, which is eliminated, if we set the initial value and the unit of measurement.

§2 Arithmetization of Space

1. Geometrical space, or simply space, is the set of things called points, lines, surfaces, angles, distances, etc., which are in certain relations among themselves that are defined by a system of axioms and theorems derived from these axioms.[29] The geometrical space can correspond to the physical material space in the sense that each entity of the geometrical space can correspond to some image of the physical space. The correspondence between the physical and the geometrical spaces can be realized by various and often fantastical methods: it is sufficient to recall the various interpretations of the so-called non-Euclidean geometries, interpretations operating with simple physical images. We will see below that the theory of relativity is also a grandiose interpretation of the geometrical space of four dimensions, an interpretation, which operates not with simple, but with highly complex physical images. It is necessary to point out here that the

---

[28]See the excellent paper: Runge, Maass und Messen. *Encyclopedie d. Mathemat. Wissenschaften*, Bd. V.

[29]See D. Hilbert, *Grundlagen der geometrie*.

physical space is a *material* space, that all images of the geometrical space are interpreted in the physical space either by *material objects* or by material actions with them.

The points of our (three-dimensional) space possess the property of having certain positions in space, that is the points can differ from one another. We can arithmetize this class of properties by assigning a given triple of numbers to each point; each triple of numbers will represent one and only one point of space. We will call such arithmetization, for brevity, *arithmetization of space*. The process of arithmetization of space is completely arbitrary; the choice of the mentioned triple of numbers is not limited by anything. Let us see how by choosing one method of arithmetization of space, we can go to any other method of arithmetization. Let in the first method a point P of space be represented by a triple of numbers x_1, x_2, x_3. Let in the second method the same point P of space be represented by another triple of numbers $\bar{x}_1, \bar{x}_2, \bar{x}_3$. Then it is obvious that we can determine the second triple of numbers from the first triple; adopting a similar definition for all points P, we will have the following relation:

$$
\begin{aligned}
\bar{x}_1 &= f_1(x_1, x_2, x_3), \\
\bar{x}_2 &= f_2(x_1, x_2, x_3), \\
\bar{x}_3 &= f_3(x_1, x_2, x_3),
\end{aligned}
\tag{3}
$$

The triple of numbers, with which we arithmetize space, are called *coordinates* (generalized); the transition from one method of arithmetization of space to another is called *a coordinate transformation*.

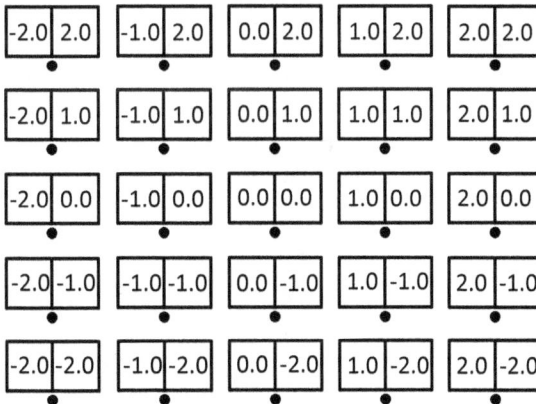

Figure 1

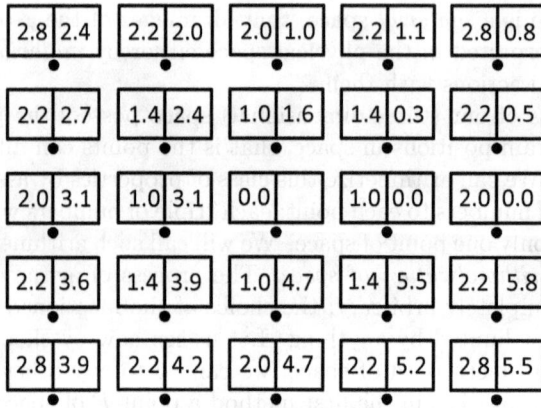

Figure 2

2. It is very important to realize, with the necessary clarity, the complete arbitrariness of the arithmetization of space in order not to confuse the arithmetization of space with the choice of a given ordinary system of coordinates, for example, orthogonal, rectilinear, polar, or any curvilinear coordinates. In fact, adopting one or another coordinate system already requires knowledge of some properties of the things that make up the space; for example, adopting orthogonal rectilinear coordinates requires knowledge of straight lines, properties of orthogonality and lengths of segments of straight lines, therefore it requires a number of axioms and the properties deduced from them. Meanwhile the arithmetization of space does not require a detailed knowledge of its properties, and can be made without having to establish in advance the nature of space. The figures (Fig. 1 and Fig. 2) depict two different methods of arithmetization of space (for simplicity, we consider two-dimensional space), where around each point we write a pair of numbers corresponding to its arithmetization by the first and the second method. It is easy, by testing, to see that the transition from the first arithmetization (x_1, x_2) to the second $(\overline{x}_1, \overline{x}_2)$ is made by the following formulas:

$$\overline{x}_1 = \sqrt{x_1^2 + x_2^2}, \quad x_1 = \overline{x}_1 \cos \overline{x}_2;$$

$$\overline{x}_2 = \arccos \frac{x_1}{\sqrt{x_1^2 + x_2^2}} = \arcsin \frac{x_2}{\sqrt{x_1^2 + x_2^2}},$$

$$x_2 = \overline{x}_1 \sin \overline{x}_2.$$

These formulas characterize the usual transition from orthogonal rectilinear coordinates to polar coordinates in the plane.[30]

3. The physical space which corresponds to the geometrical space becomes arithmetized together with the latter. The implementation of the arithmetization of the physical space around each "point" can be represented by a "board" on which we write the three numbers assigned to this point.

Upon completion arithmetization around each "point" of physical space exhibit "board" written on it with three numbers corresponding to this point. We have become so accustomed to this arithmetization of the physical space, that we do not notice at all either its essence or its arbitrariness. The names or the numbers of streets, houses, floors, apartments are themselves an example of the arithmetixation of the physical space, striking with its arbitrariness (especially in our towns and in Moscow); in the same way milestones and triangulation signs are also arithmetization of parts of the physical space around the Earth; even more complete arithmetization is reflected in the degree system of geographical maps.

It should be pointed out that the arithmetization of space has nothing to do neither with measurement nor even with estimation, since we do not define the meaning of the notions "more" or "less" referring to the location of a point in space.

4. Identifying things that make up space and clarifying their mutual relations, we can divide the properties of things in space into two categories. One will depend completely on the chosen (always at our will) arithmetization of space, the other will remain unchangeable regardless of the way we arithmetize space. Let us call the first kind of things and properties *non-intrinsic*, and the second – *intrinsic*. Studying the non-intrinsic properties of the geometrical or the physical space, we, in fact, are studying the method of arithmetization of space and only when we go to the intrinsic properties, only then we will study space as such irrespective of the arbitrariness of the adopted arithmetization. It is clear how important to us is the identification of the intrinsic things and their properties in space. One should not think, however, that the study of non-intrinsic things and properties is unnecessary ballast and error; without the knowledge of

---

[30]It is necessary to point out that the introduced method of arithmetization of the plane, using polar coordinates, has in one of the points (namely in the so-called coordinate origin) a *singularity*. Indeed, in the point $x_1 = 0$, $x_2 = 0$ the formulas giving the transition from x_1, x_2 to $\overline{x}_1$, $\overline{x}_2$ do not give a definite value for $\overline{x}_2$. The assumption which we made in Fig. 2, setting (for greater definiteness) for the given point $\overline{x}_2 = 0$, has a very significant deficiencies associated with the notion of continuity. It is impossible to discuss this issue in greater detail here.

spherical astronomy, it would be impossible to find the laws of motion of celestial bodies; moreover spherical astronomy deals precisely with the non-intrinsic properties of the physical celestial space, which is arithmetized by a given method.

The intrinsic properties of space can be expressed by propositions whose form does not change when passing from one arithmetization of space to another according to formulas (3). That is, the intrinsic properties of space are *invariant* under coordinate transformations by formulas (3). We will see later on how the invariance of the intrinsic properties of space will affect our formulas and reasoning.

5. Once space is arithmetized by a given method, it is easy to define analytically a curved surface in space. We will call a *curve* the set of points, whose coordinates are defined by

$$x_1 = \varphi_1(u), \quad x_2 = \varphi_2(u), \quad x_3 = \varphi_3(u). \tag{4}$$

We call a *surface* the set of points whose coordinates are defined by

$$x_1 = \psi_1(u, v), \quad x_2 = \psi_2(u, v), \quad x_3 = \psi_3(u, v), \tag{5}$$

where u, v are arbitrary parameters (variables). Excluding u from equation (4), we will obtain the equation of a curve in the form of two relations:

$$\Phi_1(x_1, x_2, x_3) = 0, \quad \Phi_2(x_1, x_2, x_3) = 0.$$

Excluding the parameters u, v from equation (5) we obtain the equation of a surface in the form: $\Phi(x_1, x_2, x_3) = 0$. It is clear from the above that the intersection of two surfaces is a curve and that each curve is an intersection of two surfaces.

Therefore, the definitions of the notions of a curve and of a surface need only arithmetization of space, but does not need any knowledge of the properties of space; it is also obvious that the form of equations (4) and (5) essentially depends on the method of arithmetization of space. Returning to the arithmetization of space depicted in Fig. 1 and Fig. 2, we see that the curve

$$x_1 = 1, \quad x_2 = u$$

and the curve

$$\overline{x}_1 = 1, \quad \overline{x}_2 = u$$

are completely different (in the usual geometry the first is a straight line, the second – a circle). It is not difficult to see that that property of a curve to contain a given point (to pass through the point) is an intrinsic property, exactly like the property of a surface to contain a given curve (to pass through the curve) is an intrinsic property, which is invariant under coordinate transformations.

§3 The Metric of Space

1. To any two points of space P and P' we assign a special number called *the distance between these points*. As this number is completely determined by the location of the two points P and P', i.e., by their coordinates

$$(x_1, x_2, x_3) \quad \text{and} \quad (x'_1, x'_2, x'_3)$$

the distance between P and P' is a function of these six numbers (the two triples of numbers). Denoting the number, representing the distance between P and P', by the symbol (P, P') we will have

$$(P, P') = D(x_1, x_2, x_3; x'_1, x'_2, x'_3) \tag{6}$$

and will regard the distance as an invariant property of two points; it is evident that the function D depends on the choice of coordinates, i.e., on the method of arithmetization of space. If the function D is set in a chosen arithmetization of space, we say that *the metric of space is defined*. It is clear that once the metric is defined in one coordinate system, then by formulas (3) it can be defined in any other coordinate system, i.e. by any other method of arithmetization of space.

Turning to Fig. 1 and Fig. 2, let us define in the first method of arithmetization of the two-dimensional space the distance between the points

$$P(x_1, x_2) \quad \text{and} \quad P'(x'_1, x'_2)$$

in the following way

$$(P, P') = \sqrt{(x'_1 - x_1)^2 + (x'_2 - x_2)^2}.$$

Then in the second method of arithmetization of the two-dimensional space we will have:

$$(P, P') = \sqrt{\bar{x}_1'^2 + \bar{x}_1^2 - 2\bar{x}_1' \bar{x}_1 \cos(\bar{x}_2' - \bar{x}_2)}.$$

These two formulas give the usual (Euclidean) distance between two points in orthogonal rectilinear and in polar coordinates.

2. The difficulty in determining the distance between two points depends first of all on the larger number of variables of the function

D. In order to simplify the determination of the distance, consider the distance between points P', close to point P, and point P itself, and see how the function D will be expressed in this case. We will see that when P' is close to P the expression of the distance between these points will somehow depend on the coordinates of point P and will much simpler[31] depend on very small (infinitesimally small) differences of the coordinates of the points $P'$and P. However, we will introduce in advance not the distance between points P' and P, but the square of this distance and will also use not three but only two coordinates[32]

$$(P, P')^2 = D^2 = \Delta(x_1, x_2; x_1', x_2').$$

In order to determine the function D or Δ, we, of course, need to set some restrictions, which we will impose on the notion of distance: it is clear that these restrictions will play the role of axioms in the system of geometry which we consider. Assume that the distance between the points (P, P') possess the following properties:

1) The distance does not depend on the order of points:

$$(P, P') = (P', P), \quad \Delta(x_1, x_2; x_1', x_2') = \Delta(x_1', x_2'; x_1, x_2).$$

2) The distance between two coinciding points (points with the same coordinates) is zero:

$$(P, P) = 0 \quad \Delta(x_1, x_2; x_1', x_2') = 0.$$

3) The squire of the distance between two sufficiently close points can be expanded in Taylor series in powers of the differences of the coordinates of these points:

$$
\begin{aligned}
\Delta(x_1, x_2; x_1', x_2') &= g_0 + g_1(x_1' - x_1) + g_2(x_2' - x_2) \\
&+ g_{11}(x_1' - x_1)^2 + g_{12}(x_1' - x_1)(x_2' - x_2) \\
&+ g_{21}(x_2' - x_2)(x_1' - x_1) + g_{22}(x_2' - x_2)^2 + \dots,
\end{aligned}
\tag{7}
$$

where

$$g_0, g_1, \dots, g_{11}, g_{12}, \dots$$

depend, of course, on x_1, x_2 as it should be in the Taylor series.
Since

$$\Delta(x_1, x_2; x_1', x_2') = g_0$$

---

[31]Of course, in the first approximation.
[32]This transition to two-dimensional space is necessary only to simplify the explanation.

then by formula (7) and using property 2) g_0 becomes 0. On the other hand, in the case of very small differences:

$$x_1' - x_1 \quad x_2' - x_2$$

the terms with the first powers of these quantities in equation (7) will play the main role, as long as g_1, g_2 do not become simultaneously zero, but the terms with the first powers will change their sign when we interchange the coordinates: in such a way (P, P') and (P', P) for sufficiently close points will have different signs, which contradicts property 1) of the distance; so we must assume that

$$g_1 = g_2 = 0.$$

Assuming that

$$x_1' = x_1 + \Delta x_1, \quad x_2' = x_2 + \Delta x_2,$$

and regarding

$$\Delta x_1, \quad \Delta x_2$$

as small enough, we will have:

$$\Delta = g_{11}\Delta x_1^2 + g_{12}\Delta x_1 \Delta x_2 + g_{21}\Delta x_2 \Delta x_1 + g_{22}\Delta x_2^2 + \ldots .$$

The expression for the squired distance contains four terms. But since some of them are similar we will combine them, will replace Δx_1, Δx_2 with dx_1, dx_2 (infinitesimal quantities), and will finally have:

$$\Delta = ds^2 = g_{11}dx_1^2 + 2g_{12}dx_1 dx_2 + g_{22}dx_2^2, \tag{8}$$

where $2\,g_{12}$ replaces the expression $g_{12} + g_{21}$. Therefore, g_{ik} is symmetric on its indices, i.e., $g_{ik} = g_{ki}$. It is obvious that g_{ik} depends on x_1, x_2. We will denote the quantity Δ by ds^2; so ds will be the distance between two infinitely close points (in first approximation).

Thus, the squared distance of two infinitely close points is a square function of the infinitesimal coordinate differences of these points, with coefficients depending on the coordinates of the main point. The values of these three coefficients in (8) for any of the points of the two-dimensional space completely determine the metric of the distances in our space in the immediate vicinity of the given point. It is possible to show[33] that the value of these coefficients makes it possible to define the squired distance between any two points in such a way that

---

[33] Here we cannot give the proof of this assertion.

18

the expression (8) would give the distance between two infinitely close points. Noticing that the study of distances in the physical space, in its turn, will be based on distances between points very close to each other, let us regard *the metric of the distances in space as defined when, in a given arithmetization of space, g_{ik} are given as functions of the coordinates of the points of space.*

All quantities g_{ik}, as functions of the coordinates of point P, are called a *fundamental metric tensor.*

In the case of three-dimensional space we will have not three, but sic coefficients g_{ik}, and formula (8) will obtain the form:

$$ds^2 = g_{11}dx_1^2 + g_{22}dx_2^2 + g_{33}dx_3^2 + 2g_{23}dx_2dx_3 + 2g_{31}dx_3dx_1 + 2g_{12}dx_1dx_2.$$
$$(9)$$

Thus, in three-dimensional space the fundamental metric tensor contains six functions of the coordinates of our point.

3. In view of the very import role of the fundamental metric tensor for what follows, we will consider the distances defined above by the two methods of arithmetization of two-dimensional space, mentioned earlier.

In the first method we have

$$\Delta = (x_1' - x_1)^2 + (x_2' - x_2)^2,$$

in other words:

$$g_0 = 0, \quad g_1 = 0, \quad g_2 = 0,$$
$$g_{11} = 1, \quad g_{12} = 0, \quad g_{21} = 0, \quad g_{22} = 1,$$
$$ds^2 = dx_1^2 + dx_2^2,$$

and we obtain the usual formulas of analytic geometry for the squared distance and the squared element of arc.

Somewhat more complicated are the calculations for the arithmetization by the second method; without giving these calculations, we note that for the second method of arithmetization

$$\bar{g}_0 = 0, \quad \bar{g}_1 = 0, \quad \bar{g}_2 = 0,$$
$$\bar{g}_{11} = 1, \quad \bar{g}_{12} = 0, \quad \bar{g}_{21} = 0, \quad \bar{g}_{22} = \bar{x}_1^2,$$
$$ds^2 = d\bar{x}_1^2 + x_1^2 d\bar{x}_2^2,$$

where in this formula we recognize the element of arc in polar coordinates.

4. It is completely clear that, knowing Δ or ds for any points infinitesimally differing from point P, we will know the fundamental

metric tensor at point P; we can also, using the fundamental metric tensor, determine the length of the arc of the curve between two points P_1 and P_2. We can place an infinite number of points between P_1 and P_2 on the curve and sum up the distances between them. We will call the number, representing the sum of all these distances, the length of the arc of our curve between the points P_1 and P_2. The corresponding formulas of calculus give the following expression for the length S_{P_1,P_2} of the arc of the curve of the curve between points P_1 and P_2:

$$S_{P_1,P_2} = \int_{u_1}^{u_2} \sqrt{g_{11}\left(\frac{\mathrm{d}x_1}{\mathrm{d}u}\right)^2 + g_{22}\left(\frac{\mathrm{d}x_2}{\mathrm{d}u}\right)^2 + \ldots}\; \mathrm{d}u, \qquad (10)$$

where u_1, u_2 are the values of the parameter u of our curve for the points P_1 and P_2.

5. Let us see how the study of the physical space can indicate (at least in principle) what is the dependence of the fundamental metric tensor on the coordinates of points, and therefore what is the metric of the distances of the geometric space, which corresponds to the physical space, which in turn is the interpretation of the geometric space.[34] Suppose that we are defining a special property of physical distance along a physical curve between physical "points" lying on the curve. Suppose that by using certain physical manipulations we transform this property at first into the category of intensities, and after that into measurable intensities. Recall that in order to do this we have to define the notions "more" or "less" and the notion of equidistant intensities with respect to this physical distance along a curve. Suppose further that, by a certain choice of the initial value and the units of measurement we measure the introduced intensities. The result of the measurement will be a certain number, which we call *the physical length of the arc along a given curve between the points P_1 and P_2.* If the physical length of the arc is an interpretation of the defined in Section 4 above geometrical length of the arc (or just the length of the arc), it is evident that both numbers, giving the geometrical and physical length of the arc, should coincide (at least for some choice of the initial values and the units of measurement). Since the geometrical length of the arc between two coinciding points is zero [by the property of integral (10)], it is obvious that the physical length between two coinciding points should be chosen as the initial value; the choice

---

[34]EDITOR'S NOTE: It is helpful to keep in mind that a mathematical object is interpreted by a physical object, whereas a physical object is represented by a mathematical object.

of units of measurement should not be restricted by anything.[35]

Assume now that we draw a number of curves through a given physical point P, and on each of these curves choose a point P very close to P'. Measuring the physical length of the arc along each of these curves between the point P and the one infinitesimally close to it, we obtain a series of numbers $\Delta s_1, \Delta s_2, \ldots$, which can be identified with the corresponding geometrical lengths of the arcs, and since the arcs are very small their lengths can be identified with the distances between very close points, and this distances can be calculated by formula (9), replacing dx_1, dx_2, d in it with the very small quantities $\Delta x_1, \Delta x_2, \Delta x_3$ which are differences of the coordinates of the points P' and P. In formula (9) the quantities g_{ik} will be unknown, but $\Delta x_1, \ldots$, and the left-hand side of the obtained equation will be known. By doing a sufficient number of measurements, we will be able to establish a sufficient number of equations from which g_{ik} can be determined; obviously, we will need at least six measurements in the case of the three-dimensional space, whereas in the case of a two-dimensional space – at least three.

6. To make the process of the physical study of the metric clearer, we will explain it in more detail in the case of a two-dimensional space. As we just saw we need three measurements. To do this, in addition to point P, we choose three more points – P_1, P_2, P_3 – which are very close to P. By doing measurements, we determine three distances (very short arc lengths of the curves) s_1, s_2, s_3 between point P and the points P_1, P_2, P_3. To determine the metric of space we need to arithmetize space, one way or another, so we have to choose some method of arithmetization. Let, by the chosen method, point P have coordinates (x_1, x_2), and points P_1, P_2, P_3 – $(x_1 + h_1, x_2 + k_1)$, $(x_1 +$

---

[35]This arbitrariness of the units of measurement is quite characteristic. Weyl uses it for defining a special notion of the arbitrariness of the scale. We can use an arbitrary and changing from point to point measurement unit for measuring a physical length; the question is what choice of the measurement unit will make the physical length identical with the geometrical length. If this question is left undefined (as it should be since the units of measurement depend on our will), then the physical length should be defined not be ds, but by $\mu(x_1, x_2, x_3)\, ds$, where μ is an undefined function of the coordinates. There exist properties of space which do not depend on μ. We call these properties *scale invariant*. They play important role in Weyl's theory. However, we will not discuss here the question of scale invariance, because it will overload this work with new notions and terminology, which will lack concrete content since they will not be supported either by a mathematical apparatus or examples.

$h_2, x_2 + k_2), (x_1 + h_3, x_2 + k_3)$. The formula (8) gives three equations

$$s_1 = g_{11} h_1^2 + 2 g_{12} h_1 k_1 + g_{22} k_1^2$$
$$s_1 = g_{11} h_2^2 + 2 g_{12} h_2 k_2 + g_{22} k_2^2$$
$$s_1 = g_{11} h_3^2 + 2 g_{12} h_3 k_3 + g_{22} k_3^2.$$

We can determine the metric of space by finding g_{11}, g_{12}, g_{22} from this system of equations. The solution of this system of equations is always possible, if the points P, P_1, P_2, P_3 do not coincide; unfortunately, I cannot elaborate further on this point here. What was now described with sufficient pedantry, occurred in Egypt, where practical life and religious needs established the foundation of the Euclidean metric. Imagine for a moment that we are unnaturally flattened and like surface shadows live on the big sphere;[36] suppose that by going to the two-dimensional existence of shadows, we still know how to measure the physical lengths of arcs; suppose that for the arithmetization of our two-dimensional space (the spherical surface), we use numbers as follows: introduce the usual methods of addition to 90^0 of latitude φ and longitude λ on our sphere of radius R, then to each point P assign two numbers x_1, x_2 by the formulas $x_1 = R\varphi$, $x_2 = R \sin \varphi_0 \lambda$, where φ_0 is the addition to 90^0 of the latitude of a given point $P_0 (x_{10} = R \varphi_0)$, in whose vicinity we carry out our observations. Suppose that by a large number of accurate measurements we obtain the following expression for ds^2:

$$ds^2 = dx_1^2 + \frac{\sin^2(x_1/R)}{\sin^2(x_{10}/R)} dx_2^2.$$

It is easy to see that this is the usual formula of the length of an infinitesimally small arc of the sphere:

$$ds^2 = R^2 d\varphi^2 + R^2 \sin^2 \varphi \, d\lambda^2.$$

If the radius of the sphere is small compared to the area inhabited by shadows, the resulting formula will soon, after a sufficiently large number of measurements at different points in the world of shadows, give the geometers there the opportunity to determine R, i.e., the radius of the sphere, and therefore to move up significantly in the study of the two-dimensional space which is occupied by these shadows. But the task of our shadows will become more difficult, if the area occupied by them is very small with respect to the radius R of the sphere. Then

---

[36]This comparison was used by professor O. D. Hvolson in one of his, usually brilliant, popular lectures on the principle of relativity.

the ratio

$$\frac{\sin^2(x_1/R)}{\sin^2(x_{10}/R)}$$

will be close to 1, because x_1 will be close to x_{10}, i.e., $x_1 - x_{10}$ will be very small compared to R, and then everything will be as if $R = \infty$, and the sphere will turn into a plane. Our spherical shadows will tend to persecute freethinkers who doubt that their space is not a plane but a sphere of a very large radius, and will take many "spherical" centuries for the crowds of shadows to realize the sphericity of their world. We, three-dimensional beings, are in a position similar to the two-dimensional shadows, living on a sphere of a very large radius, because all our measurements continually convince us in the excellent agreement of the metric of our space with the Euclidean metric, and we need huge astronomical distances or those ideas introduced by the theory of relativity, in order that we put into question the metric of our space.

7. So, at first glance, it seems that by exploring our physical space and the physical length, we can, without any arbitrariness, determine uniquely the metric of that geometrical space, whose picture (interpretation) is the studied physical space. But we should not lose sight of a number of arbitrary conventions introduced by us when we, on the one hand, interpret the geometrical space and its images by using images of the physical space, and, on the other hand, when we identify the physical length with the geometrical length.

In the physical material space we interpret geometrical points by various material objects. Satisfied with rough interpretations, we "represent" points by small material bodies located in the physical space (compare points on paper, survey markers on the Earth's surface and, finally, for large areas of space – the heavenly bodies); interpreting more finely, we define a point as the intersection of two sufficiently narrow light beams, i.e., also by thinner but still material objects. We can, however, interpret geometrical points by entirely different images of the material world and in such an interpretation we arrive, even in the narrow accessible areas of the physical space, at entirely different ideas about its geometrical properties, or more correctly – at the properties of the geometrical space, which is interpreted by the physical space.

The determination of the physical length involves even greater arbitrariness. Here, if we do not adhere to the point of view of reasonableness, there exists a variety of possible methods for the determination of the physical length, and thus there exists a variety of determinations of the metric of that space, whose interpretation is our physical space.

Interpreting the geometrical point not in the usual way in our physical space, and establishing a special notion of physical length, we arrive experimentally, without any difficulty, to the fact that the metric of our space is not Euclidean metric but is Lobachevsky metric or some other.

Thus, we must reject a possibility of an unambiguous and independent of our will answer to the question about the metric (and, of course, about other properties) of the geometrical space, which corresponds to our physical space. This question makes sense at all, only after adopting an interpretation of the things of geometrical space by images of physical material objects, and likewise after the definition of physical length and making an assumption of its identity with the geometrical length. Once this is set, then only the question of determining the metric of space remains. The task of the geometer-naturalist consists of designing more accurate methods for studying the metric of space, which provide less rough methods, than those described in the preceding sections, for determining the metric of space. We will see how the principle of relativity by considering universal gravitation made it possible to reach some conclusions about the metric of space on the basis of so subtle phenomena as the insignificant and the unexplained motion the perihelion of Mercury, or the slight deviation of a light ray passing near the solar surface.

Naturally, a question arises, what determines the choice of one or another physical interpretation of the space and the things in it? Here probably the dark principles of reasonableness or economy of thought play a primary role – I will not even try to discuss these supernatural issues, moreover they are not necessary for the subject of this work. We adopt that interpretation of of geometry with which physicists are accustomed.[37] Touching upon the issue of physical space, it seems to me useful to draw attention to the fact that the physical space as a material space, by its very nature, is unthinkable without matter; an empty physical space is just nonsense, because in such a space it would not be possible to represent or interpret any of the things of the geometrical space – because we have agreed to interpret the things of the geometrical space by *material images* of the physical space.[38]

---

[37]For example, a very small material body interprets (with a certain approximation) a point of the geometrical space; a light ray interprets a straight line, etc.

[38]EDITOR'S NOTE: This sentence is puzzling since Friedmann was well aware of de Sitter's solution of Einstein's equations which describes a matter-free universe – see his paper "On the Curvature of Space" in Alexander A. Friedmann, *Papers On Curved Spaces and Cosmology* (Minkowski Institute Press, Montreal 2014).

§4 The Curvature of Space

1. Embarking on the presentation of one of the most difficult geometrical notions, namely the notion of curvature of space, I consider
it necessary to warn you that in this section we present only the most
basic foundations of the theory of curvature since a more thorough
discussion of the theory of the curvature of space requires much more
heavy mathematical apparatus than the one we use in this work.

Drawing different curves through a given point P of the three-
dimensional space, we associate with each curve a special notion of a
direction of this curve at the point P. Selecting on a curve a point
P' which is infinitely close to the point P, we define the direction
of the curve by two numbers characterizing the ratio of infinitesimal
increments of coordinates occurring when we move from point P to its
infinitely close point P'.

Consider, for simplicity, the case of a two-dimensional space and
assign the coordinates x_1, x_2 to point P, and the coordinates $x_1 + dx_1$,
$x_2 + dx_2$ – to point P', we define the direction of the curve by the
ratio: dx_2/dx_1.

Obviously, the direction will be determined as long as we are given
a point P and its infinitely near point P', i.e., as long as we know
the difference of the coordinates of these points. In three-dimensional
space such differences will be three – dx_1, dx_2, dx_3; in two-dimensional
space – two (dx_1, dx_2). Let us call the thing, which is well-defined by
these two points or the differences of their coordinates, an *infinitesimally small vector at point P*; we will call the coordinate differences
themselves the *components of the infinitesimally small vector at point
P*. In the case of our usual geometry, an infinitely small vector can be
interpreted as a line segment from P to P', and its components are
then the projections of this segment on the axes of orthogonal rectilinear coordinates. From the above it is clear that the direction of a curve
at a point P is well defined once we define the corresponding infinitely
small vector at point P. Thus, instead of talking about the direction
of a curve at the point P, we can now talk about the direction of an
infinitesimal vector at point P.

2. The directions of any two curves passing through a point P,
define themselves a new notion of the angle between these directions.
Since the directions of the curves coincide with the directions of the
two corresponding infinitesimal vectors, then we can talk about *the
angle between two infinitesimal vectors at point P*. This angle will be
a certain number, which is determined by the components of both infinitesimal vectors and by the values of the fundamental metric tensor

at point P. To explain how this number is determined, we turn again to the two-dimensional space. Suppose first that the two-dimensional space is the usual Euclidean plane, and its arithmetization is produced by introducing orthogonal rectilinear coordinates (as in Fig. 1). Then the distance of two infinitely close points P and P' is determined by the formula:

$$ds^2 = dx_1^2 + dx_2^2.$$

Thus, for the fundamental metric tensor in this case we will have:

$$g_{11} = 1, \quad g_{12} = g_{21} = 0, \quad g_{22} = 1.$$

Now take (see Fig. 3) two infinitesimal vectors (dx_1, dx_2) and $(\delta x_1, \delta x_2)$, depicted by two segments PP' and PP''. By the rules of analytical geometry the angle ω between these two vectors is determined as a certain number according to the following formula:

$$\cos \omega = \frac{dx_1\, \delta x_1 + dx_2\, \delta x_2}{\sqrt{dx_1^2 + dx_2^2}\, \sqrt{\delta x_1^2 + \delta x_2^2}}.$$

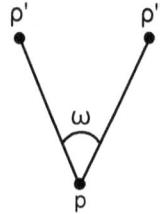

Figure 3

Turning to another arithmetization of the Euclidean plane or considering, from a more general point of view, a two-dimensional space with properties, other than those of the Euclidean plane, let us define the angle between two vectors (dx_1, dx_2) and $(\delta x_1, \delta x_2)$ as a number obtained by the formula:

$$\cos \omega =$$
$$= \frac{g_{11}\, dx_1\, \delta x_1 + g_{12}(dx_1\, \delta x_2 + dx_2\, \delta x_1) + g_{22}\, dx_2\, \delta x_2}{\sqrt{g_{11}\, dx_1^2 + 2\, g_{12}\, dx_1\, dx_2 + g_{22}\, dx_2^2}\, \sqrt{g_{11}\, \delta x_1^2 + 2\, g_{12}\, \delta x_1 \delta x_2 + g_{22}\, \delta x_2^2}}.$$

This formula shows that two vectors at a point P will form an angle different from the angle, which these same vectors (with the previous components) form at another point P_0, because $g_{11}, g_{12}, \ldots$ depend on the coordinates of point P and therefore change when we go from point P to point P_0. From this observation it is evident that a space with different properties than the Euclidean (with a different metric) differ radically from that space, with which we are accustomed to operate. We can prove that the angle between two infinitesimal vectors, which we defined, does not depend at all on the method of arithmetization of space.

3. The role of the Euclidean parallel postulate in the development of geometrical ideas, especially in the XIX and XX centuries,

is well-known. The notion of parallelism plays a huge role in spaces whose metric properties are different from those of Euclidean space; it should, however, be suitably generalized and defined. Such a generalization of the notion of parallelism was realized in the idea of *parallel transport* developed by the Italian mathematician Levi-Civita.

Imagine that an infinitesimal vector (Fig. 4) moves along a curve and changes according to some law. The law of this change must be such that, by specifying an infinitesimal vector at a point

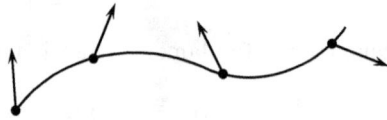

Figure 4

of the curve, we can determine a vector, corresponding to the first, at any other point of the curve; further, the law of this change should enable us to carry out that transport of the vector along any curve in space. Let us call the just described manipulation of vectors *conjugation*. We will call a specifically chosen conjunction *parallel transport* of a vector. According to the way we choose this conjugation we will have certain geometrical properties of our space: in other words, certain parallel transport will define a special class of space.

It is possible, as I already said, to choose the parallel transport arbitrarily; however, we will concentrate on parallel transport, which is the closest to our usual notions, namely, on such *parallel transport, which preserves the angle between two parallel transported vectors along the same curve* (Fig. 5).[39]

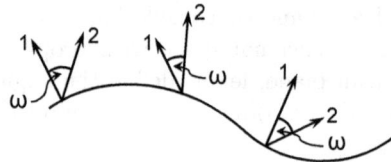

Figure 5

The distance between two infinitely close points P and P', defining a vector, depends, as we have seen, on the fundamental metric tensor and on the components of our vector; we will call this distance the *magnitude of the vector*. The magnitude of the vector generally changes, when it is parallel transported, and only in very exceptional cases remains unchanged. Let us call the spaces in which the magnitude of a vector does not change under parallel transport *Riemannian*[40] *spaces*, whereas the spaces where the magnitude of a

---

[39]Considering space, we must choose the parallel transport in a given way; the different choices of parallel transport will define a number of properties of our space.

[40]A space of constant Riemannian curvature is a special kind of Riemannian

vector changes under parallel transport – *Weyl's spaces*. The Riemannian spaces have an extremely curious feature: in them the parallel transport is completely defined by the knowledge of the metric of space (i.e., by the knowledge of the fundamental metric tensor). The case of Weyl's spaces is more complex: here, in addition to the knowledge of the fundamental metric tensor, the definition of parallel transport also requires knowledge of some more quantities; the set of these quantities is called a *scale vector*. The scale vector in three-dimensional space consists of three quantities, in two-dimensional – two.[41]

To give a clearer idea of parallel transport, consider this notion in the ordinary Euclidean plane by defining the parallel transport of a vector as an ordinary parallel transport of a vector from one point to another (Fig. 6), a transport which does not change the magnitude of the

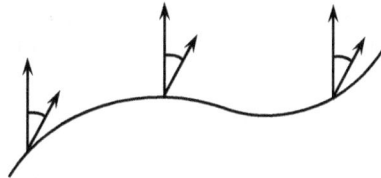

Figure 6

vector, and the straight lines, on which the vector is lying before and after the transport, remain parallel (in the usual sense). It is evident that such a parallel transport of a vector can be defined analytically by the following method: the vector (dx_1, dx_2) is parallel transported if the vector $(\delta x_1, \delta x_2)$ after the parallel transport is defined by:

$$\delta x_1 = dx_1, \quad \delta x_2 = dx_2.$$

It is possible to define a notion of parallelism from this definition of parallel transport and introduce a number of properties corresponding to this notion in Euclidean geometry. It is easy to see that, with the above definition of parallel transport, the angle between two parallel transported vectors does not change, as well as the magnitude of a parallel transported vector does not change either. So the Euclidean parallel transport characterizes a space, which is a special case of Riemannian space.

4. The parallel transport makes it possible to define the notion of a straight line. At each point of a curve, this curve has a definite direction, characterized by the direction of an infinitesimal vector. Consider an infinitesimal vector at some point P_0 of the curve,

space.

[41] Here we cannot discuss the notion of scale vector in greater detail.

which defines the direction of the curve at this point (Fig. 7). Let this infinitesimal vector be parallel transported to a point P of the curve. Then, in most cases, the direction of the vector transported to point P will not coincide with the direction of the curve at this point. A curve possessing the exclusive property that the direction of a parallel transported vector at any point P coincides with the direction of the curve at that point, will be called a *straight line*. The straight line differs from all other curves by the exceptional property that its direction along the line itself is parallel transported. Needless to say that for different definitions of parallel transport we will have different definitions of straight lines. We now turn to the Euclidean plane. There, there will obviously be curves which do not have the property that their direction along the curve is parallel transported (Fig. 8), but there will be also straight lines as defined above and such "straight lines" will coincide with the usual definition of the Euclidean straight line.

Figure 7

In Riemannian spaces the straight line defined by parallel transport has the characteristic property of being the shortest line connecting any two points located on it; in other words, the length of the arc along a straight line from P_1 to P_2 is shorter than the length of the arc along any other curve. In Fig. 9 I purposely depicted the "straight line" not as a usual straight line, keeping in mind that this drawing depicts not the Euclidean plane, and therefore its metric is so different from the Euclidean, that the length of the arc $P_1 P_2$ along the "straight line" A is shorter than the length of the arc $P_1 P_2$ (in the new metric) along the Euclidean straight line B.

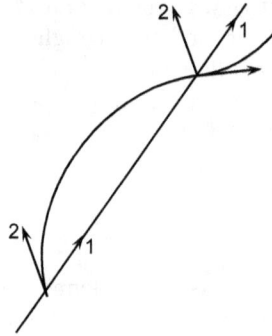

Figure 8

So, *the straight lines of the Riemannian spaces are the lines of the shortest distances* (geodesics); the straight lines of Weyl's spaces do not possess this property.

With the help of the definition of a straight line, it is easy to parallel transport any vector along a straight line, at least in a Riemannian space; an infinitesimal vector of direction, by definition, is parallel transported along a straight line.

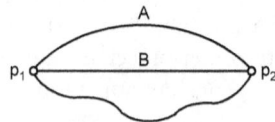

Figure 9

The angle between two parallel transported vectors, as we know,
does not change, and therefore in or-

Figure 10

der to parallel transport a vector along a straight line, it is sufficient
to monitor that it always forms the same angle with the direction of
the straight line and that its magnitude does not change. Fig. 10 and
Fig. 11 depict parallel transport of a vector, where in Fig. 10 as a
straight line is taken the usual straight line, whereas in Fig. 11 – some
curve.

On the Euclidean plane parallel
transport along a usual straight line
characterizes by the fact that the transported vector will be, in the usual sense,
parallel to its initial direction (see Fig.
10). As is known, the arc of a great circle is the shortest line on a sphere. Fig.

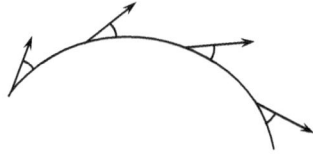

Figure 11

12 shows parallel transport of an infinitesimal vector along the arc of
a great circle. Let us repeat: the angle ω_1 between the vector and the
arc at the initial point is equal to the angle ω_2 between the parallel
transported vector and the arc at the final point.

5.   Let us draw any closed line
through a given point P – the contour
C (Fig. 13), take some infinitesimal vector a at point P and parallel transport
it along the line C until we return to
the starting point P; vector a will become vector a_1 and its magnitude (in
Weyl's space) and its direction (in both
Weyl's space and Riemannian space), in
general, will not coincide with the magnitude and direction of vector a_1. The
case when a_1 coincides with a by both
magnitude and direction will be an exceptional case.

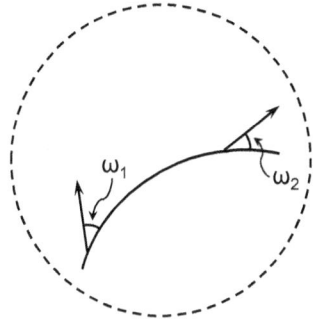

Figure 12

Let us denote by α the angle which a_1 forms
with a. On what quantities may, in general,
this angle depend? Obviously, the angle will
depend on the properties of the given space
(because the parallel transport itself characterizes space) and on the shape of the closed
line C. Drawing shrinking closed lines through
point P, we notice that the shape of the closed

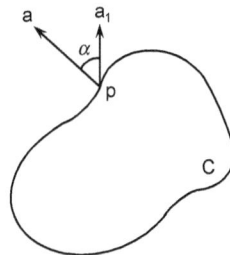

Figure 13

line will affect less and less the angle α; in two-dimensional space α is proportional to the area s[42] bounded by the closed curve, and the coefficient of proportionality will depend on the location of point P. This coefficient of proportionality is called *vectorial curvature or just curvature of the two-dimensional space at point* P; denoting the curvature by K we have

$$K = \lim_{s \to 0} \frac{a}{s}. \tag{11}$$

The curvature in three-dimensional space is defined analogously; it would be too difficult to discuss the definition of curvature in three-dimensional space in detail here. It is sufficient to know, that the angle α (between the vector, which returned to the starting point after the parallel transport, and the vector at the starting point) characterizes the *mean curvature* of the area of space where the closed curve is located. The greater this angle, the greater the mean curvature. If this angle is zero, it means that the average curvature of space of its part under consideration is zero.

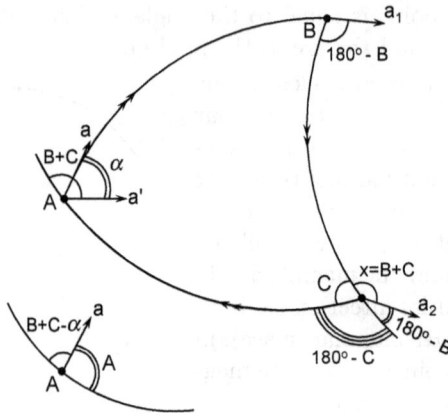

Figure 14

To give a visual representation of curvature, imagine again that we are two-dimensional shadows living on a sphere of radius R. On this sphere we draw a spherical triangle ABC consisting of three "straight"

[42]Of course, the notion "area" should be defined. Without going into specific details, we note that for two-dimensional space the area of the space bounded by a closed curve will be a number defined by the integral $\iint \sqrt{g}\, dx_1 dx_2$, where $g = g_{11}g_{22} - g_{12}^2$, and the integral is taken over the area bounded by our closed curve. For the Euclidean plane the notion of area defined in this way coincides with the usual notion; it is the same for the area of figures on the sphere.

lines (Fig. 14), i.e., consisting of the arcs of three great circles (the two-dimensional space on the sphere is a Riemannian space). Let A, B, C denote three internal angles of our triangle. At point A take an infinitesimal vector a whose direction coincides with the direction of the arc AB (a "straight" line). Then parallel transport this vector along the closed line forming the triangle ABC; our vector will return at the starting point A as a vector a' with already changed direction.

Let us calculate angle α between the vectors a and a'. For this purpose we will successively parallel transport the vector a along the arc AB, along the arc BC and along the arc CA. Since AB is a "straight" line, during its parallel transport to point B the vector a will keep the direction of the arc AB, and when it arrives at point B as the vector a_1, will have the direction of the arc AB at B, and therefore will form an angle of $180^0 - B$ with the arc BC. The parallel transport of vector a_1 along the "straight" line (the arc) BC, according to subsection 4 above, will preserve the angle between a_1 and the "straight" line BC; therefore the angle between vector a_2 (resulting from the parallel transport of a_1 to point C) and the arc BC will be the same as the angle between vector a_1 and the arc BC, i.e., will be equal to $180^0 - B$. Let us now determine the angle x between the vector a_2 and the arc CA, along which it will be parallel transported. It is seen in Fig. 14 that the sum

$$x + (180^0 - B) + (180^0 - C)$$

forms an angle of 360^0 and therefore

$$x - B - C + 360^0 = 360^0 \quad \text{and} \quad x = B + C.$$

Thus, the angle between a_2 and the arc CA will be $B + C$. This angle is preserved when a_2 is parallel transported along the arc of the "straight" line CA and becomes vector a' at point A. Hence a' will form an angle with the arc CA equal to $B+C$, if α is the angle between a and a'; then, as seen in Fig. 14, the angle between vector a and the arc CA is $B+C-\alpha$. On the other hand, the angle between a and the arc AC is A. The sum of these angles is equal to 180^0 and therefore

$$B + C - \alpha + A = 180^0,$$

$$\alpha = A + B + C - 180^0.$$

In other words, the angle α is the so-called spherical excess which is equal, as is known from spherical trigonometry, to the ratio of the area of the spherical triangle ABC and the square of the radius of the sphere

$$\alpha = \frac{s}{R^2}, \quad \frac{\alpha}{s} = \frac{1}{R^2}.$$

Thus, by the definition of the curvature K we will have

$$K = \frac{1}{R^2},$$

i.e., the curvature of the two-dimensional spherical space is inversely proportional to the square of the radius of the sphere. By repeating the same procedure with a flat triangle, we would find that the angle $\alpha = 0$, because the sum of the angles in the flat triangle is equal to two right angles (180^0), and as a result we would find that the curvature of the plane equals zero.

The above procedure shows the important connection of the curvature with the sum of the angles of a triangle, and, as is well-known, the sum of the angles of a triangle is closely connected with Euclid's postulates; unfortunately, however, I cannot discuss this issue in more detail. The study which the two-dimensional spherical beings have just completed, would enable them to determine the curvature of their space, and consequently the radius of the sphere, by which they could interpret the geometry of their world. We will again return to this remark later on.

6. When a vector is parallel transported along a closed curve C from point P, it will return at the same point as the vector a' and, in general, its magnitude will change. For Riemannian spaces, as already indicated above, the magnitude of a vector does not change under parallel transport, so the magnitude of a' will be equal to the magnitude of a. In the case of Weyl's spaces the magnitude of a' will be different from the magnitude of a. Let us examine this change of the magnitude of a parallel transported vector. It will depend on three factors: first, on the initial magnitude of the vector, second, on the type of the closed curve C and, third, on the properties of our space. Let l, l' be the magnitudes of the vectors a and a'; $\Delta l = l' - l$ will be the change of the magnitude of vector a under its parallel transport along a closed curve; $\lambda = \Delta l/l$ *will be the relative change of the magnitude of vector a under its parallel transport along the closed curve C.*

Let us turn again to the two-dimensional space; the relative change λ of the magnitude of vector a will be, if C is sufficiently small, proportional to the area s, bounded by the curve C, wherein the coefficient of proportionality will vary depending on the position of the starting point P. This coefficient of proportionality is called by Weyl *metrical curvature of the two-dimensional space at point* P; denoting the metrical curvature by L we will have:

$$L = \lim_{s \to 0} \frac{\lambda}{s}. \tag{12}$$

The metrical curvature in three-dimensional space is defined is a similar way. For us, it is sufficient to know that the relative change of the magnitude of a vector under parallel transport characterizes the *mean metrical curvature* of the area of space where the closed curve C is located. It is clear that the metrical curvature of Riemannian space is zero, but it should be pointed out that the metrical curvature can be also zero for non-Riemannian spaces.

Both the vectorial and the metric curvatures do not change from point to point in space. The vectorial curvature is completely determined by the fundamental metric tensor and the scale vector, whereas for the computing of the metric curvature only one scale vector is required. Obviously, both the vectorial and the metric curvatures are intrinsic properties of space and do not depend on the method of its arithmetization. All the time when we talked about space, we meant three-dimensional space and turned to two-dimensional space for illustrative purposes only. It is easy, however, to generalize the definitions of our notions and their properties to a space of any dimensions. Formally, we will not encounter any difficulties, except for increasing the number of components g_{ik} of the fundamental metric tensor and the number of components of the scale vector. For two-dimensional space we have three quantities for determining the fundamental metric tensor and two quantities for determining the scale vector. In the (usual) three-dimensional space we have six quantities for determining the fundamental metric tensor and three quantities for determining the scale vector. Finally, in four-dimensional space the number of quantities needed for determining the metric tensor increase to ten (the number of different combinations of indices 1, 2, 3, 4 by two), and the number of quantities necessary for determining the scale vector reaches four.

7. Let us now turn again to the physical space and see how we can, with the help of physical manipulations, determine the curvature of the geometrical space, whose interpretation is the physical space. For this purpose it is necessary, first of all, to identify the things of the physical space, which correspond to the notions of direction, vector, and angle in the geometrical space. Since the physical space is a material space, then these things should be associated with certain material objects, in other words, the direction should be interpreted by mate-

rial bodies or processes (for example, using a light beam), the angle must be obtained as a result of a certain manipulation with material bodies (physical measurement of the angle). Further, for determining the curvature it is necessary to identify a process in the physical space corresponding to the parallel transport of a vector. Having identified this process, we can realize the parallel transport of a vector with the help of manipulations of material bodies and we can also determine its magnitude and the angles between the vector and other vectors or directions. Having identified these notions and actions in the physical space, we can perform a series of experiments to study the curvature of our space: outline a number of closed paths and start carrying out (in fact, material) parallel transport of vectors along these paths; returning to the starting point we can measure the angle α at which our vector deviated from its initial position, and determine the relative elongation λ of our vector; knowing these values for a large number of closed lines, we can determine the mean vectorial and metric curvatures of space at its various locations.

Let us assume that we would be able to perform (at least in principle) this experiment; below I will point out that this experiment, as well as the determination of the fundamental metric tensor in three-dimensional space, cannot be performed (even in thought). For small closed lines the result in our space would be rather doubtful, because the angles α and the quantities λ would be very small and it would be practically impossible to measure them. Gauss' famous measurement of the sum of the angles of a triangle is, as we saw above, the realization of an experiment to determine the angle α. The triangle had been taken on the Earth's surface, hence it had a relatively small area, and the result was such a small quantity α that it could not be measured. Things could, of course, change if as a closed line we would take a loop extending from our solar system to the Andromeda nebula, the area of such a loop would be very large, and no matter how small the curvature of our space were, the angle α and the magnitude λ would be so significant that they would be easily measured; performing these measurements, connecting them with the measurements necessary to determine the metric of space, we could easily determine the properties of the geometrical space, whose interpretation, *with certain assumptions and arbitrary* (not from the viewpoint of reasonableness) *conditions*, is our physical space. Keeping in mind the reasonableness of the chosen interpretation, we could simply say that our physical space possesses such and such properties. In other words, we would create the *physics* of space or, if you like, *physical geometry*.

However, as I mentioned above, the implementation of such experiments is quite unthinkable, and the reason for this is the fact that we cannot perform any physical action in three-dimensional space, because we have to perform all these actions in time as well, because all living things, which are able to perform physical actions live in time: παντα ρει και ουδεν μενει.[43]

There is no physical phenomenon that can happen instantly; we cannot transport a vector along a curve with an infinite speed (instantly), which means that when a vector returns back to the starting point, and we can measure the angle between it and the initial vector, it is not our old angle α, but something completely new, because for the time our vector traveled to Andromeda and back, many tens of thousands of years passed, no matter how quickly our vector was transported, and therefore both our vectors – the initial and the returning – managed to age significantly.

One could object to this and argue that by reducing the length of the path of the traveling vector, its aging can be reduced if it moves at great speed. Absolutely true, Gauss' experiment of measuring the angles of a triangle was performed in this way: the angles were measured (in principle, of course) almost instantly, because the experiment used light beams encircling the entire triangle in small fractions of a second; but shortening the length of the path will now reduce the area of the triangle, and this, in turn, will reduce α and λ to such an extent that they cannot be measured.

So, we absolutely cannot perform physical actions necessary for determining experimentally the physical geometry in three-dimensional space; for us these actions are as impossible as performing physical actions in two-dimensional space, where we cannot put our instruments and where we can not fit ourselves. The cause of these difficulties is *time*, without which there is no space and which is intrinsically linked not to a physical three-dimensional space, but to a physical four-dimensional space – the world. We now turn to the study of time.

---

[43] EDITOR's NOTE: *Everything flows, nothing stands still* – this is the essence of Heraclitus' teaching as presented by Plato in his dialogue Cratylus.

TIME AND THE WORLD

Time is the number of motion.
Aristotle

*Von Stund' an sollen Raum für sich und Zeit für sich völlig
zu Schatten herabsinken und nur noch eine Art Union der
beiden soll Selbständigkeit bewahren.*

H. Minkowski, "Raum und Zeit"[44]

§5 Time

1. We started the chapter on space with the definition of abstract
space. We could also proceed in the same way with time, saying that
time is a set of things called moments which are in certain relations
among themselves and with the three-dimensional space. However, I
prefer another way, starting with the examination of physical time;
the reason for this is that we have much clearer ideas of space, rather
than ideas of time. In the physical space we encounter a special phenomenon called *motion*, whose essence is that a physical (material)
point changes its position in three-dimensional space; in other words,
changes the three numbers x_1, x_2, x_3, which, in the arithmetization
of three-dimensional space, serve as its coordinates. The geometrical
representation of the positions of a material point will be a certain
curve called the *trajectory of a material point*; if the point moves in
a straight line, we will have a straight trajectory, if the point moves
along the arc of a circle, we will have a circular trajectory.

---

[44]From now onwards space by itself and time by itself will recede completely
to become mere shadows and only a type of union of the two will still stand
independently on its own. H. Minkowski, Space and Time, in H. Minkowski, *Space
and Time: Minkowski's Papers on Relativity*, edited by V. Petkov (Minkowski
Institute Press, Montreal 2012).

By choosing a trajectory, we do
not yet set the position of a material
point on it, but if we select a given
starting point on the trajectory and
from it we measure the length of arcs

Figure 15

along the trajectory with positive numbers in one direction and negative in the other (Fig. 15), then by choosing a trajectory and a number
t corresponding to the length of the arc, we completely define the position of a material point moving along a trajectory. We now associate
with each physical point M of space a certain *basic motion* and call a
clock at a given point M the instrument, showing the length t of the
arcs, traveled by a material point along the trajectory of basic motion.
It should always be kept in mind that the choice of the basic motion is
completely arbitrary, as well as completely arbitrary is the choice of a
starting point on the trajectory of the basic motion, i.e., the number
shown by the clock at the given point M is a completely arbitrary
value, which depends on our choice. On the other hand, we should
keep in mind that we operate all the time with the physical space and
physical phenomena, occurring in it, so that the clocks, we established,
can be (at least in theory) really built with material objects.

The quantity t, which is shown by the clock at a given point M is
called *physical local time at point* M.

2. Before responding to the natural and always made objection of
the non-uniformity of the introduced by us motion and its unsuitability
for our clocks, we will give several examples of local physical time.

First of all we consider *sidereal time*.[45] For all points of three-
dimensional space we choose one basic motion; so that the clocks at
all points in space will be the same. For basic motion we will take the
motion of the end of the arrow of a certain length, directed from the
center of the Earth toward a given star. The sidereal time $t_\star$ is the
length of the path, described by the end of the arrow. Sidereal time
$t_\star$ will be the same at all points of space, it will be *universal time*.
This, moreover, will be the most convenient time because a very large
number of motions will be performed in such a way, that the lengths
of the arcs traveled by a point along a trajectory will be proportional
to the difference between the sidereal times

$$s = a\,t_\star + b,$$

where a and b are constants. Such motions will be called *uniform with
respect to the sidereal time*. I repeat, a very large number of motions

---

[45]Here we use the term "sidereal time" not in the sense in which it is used in
astronomy.

will be either uniform or practically uniform with respect to the sidereal time. Therefore, the study of motion is considerably simplified when as time we select the sidereal time. It is exactly this time that was first selected as time by man and the choice of namely this time was historical, of course, absolutely necessary and natural in view of the enormous religious and mystical impression that sidereal time and its majestic motion had created and even now still impresses the human soul. The choice of sidereal time as universal time had a tremendous impact on the history of culture and, of course, on the creation of the basic laws of mechanics, as it will be discussed below. This choice, which from our point of view is completely arbitrary, seemed to be the only true and sacred, and indeed sidereal time, endowed with a number of mystical properties, made time in general mysterious and difficult to understand. Of course, not every motion is uniform with respect to the sidereal time; falling bodies near the Earth's surface or, better said, the fall of bodies in a constant gravitational field is an example of non-uniform (uniformly accelerated), with respect to the sidereal time, motion. The length of the arc traveled by the point in this motion, is expressed as depending on the sidereal time, by the following formula:

$$s = a\,t_\star^2 + b,$$

where a and b are constants, and a depends on the acceleration due to gravity: $a = g/2$ where g is the acceleration due to gravity in constant gravitational field.

3. Consider now another time, which we will, for brevity, call *gravitational time*. The basic motion, for this choice of time as well, will be the same for all points, and, hence, the clocks will be also the same, and the gravitational time will be universal. Suppose that a material point falls in a constant gravitational field and choose this motion as basic; the clocks will show the length of the path t_g, traveled by this point. This quantity will be the gravitational time. Compare gravitational time with sidereal time; for this it is sufficient to replace s with t_g in the previous formula:

$$t_g = a\,t + b,$$

from where

$$t_\star = \sqrt{\frac{t_g - b}{a}};$$

and therefore, if s is the path traveled by the end of the arrow of our clock, measuring sidereal time, then

$$s = \sqrt{\frac{t_g - b}{a}}.$$

Thus, in relation to the gravitational time the stars are moving non-uniformly; the arrows of our clocks are also moving non-uniformly, but a heavy shell falls uniformly, although differently at different latitudes of the Earth. Needless to say, that by adopting the gravitational time as true time, we would have to radically restructure all mechanics; mechanics would become, due to such a choice much more complicated, since most of the simplest and frequentest motions would turn to be non-uniform and governed by relatively complex laws. Of course, the introduction of gravitational time would be inappropriate, but, besides the principle of reasonableness and convenience, the gravitational time could be equally chosen to serve as time as the sidereal time.

Finally, we introduce a third kind of time – *pendulum time*. We construct a large number of identical pendulum clocks and take as basic motion, for any point on the Earth's surface, the motion of the end of the seconds arrow of the pendulum clocks placed at this point. The path traversed by the end of the seconds arrow of our pendulum clock from some initial point, is denoted by t_p and called *pendulum time*. For each point on the Earth's surface the pendulum time will be proportional to the sidereal time, so that motion which is uniform with respect to the sidereal time will be also uniform with respect to the pendulum time, and vice versa. But unlike universal time or gravitational time, the pendulum time will be local and at different latitudes will be different.

4. The above also clarifies the ever made claim about the advantages of the sidereal time, as if the sidereal time is based on uniform motion. In fact, the notion of uniformity of motion already presupposes the existence of time, and the expression "the stars move uniformly" only means what we *call* the motion of the stars uniform. The uniformity of motion is an entirely relative concept[46] – we can talk about a uniform motion with respect to another, and when we talk about uniform motion, we mean that that motion is uniform relative to

---

[46]EDITOR'S NOTE: This assertion is a bit puzzling given that Friedmann had excellent understanding of Minkowski's spacetime formulation of Einstein's special relativity. Minkowski stressed the absolute nature of both uniform and accelerated motion (revealed by his revolutionary discovery that the world is four-dimensional with one time and three space dimensions) – a uniformly moving particle is a *straight* worldline, whereas an accelerating particle is a *curved* worldline. There is no relativity in spacetime.

the motion of stars, which sounds even more odious, if we say uniform with respect to the Earth's rotation. Behind the mystic significance of the sidereal time appears to be man's unwillingness to understand the not central and modest position of the planet, on which, by the will of fate, he had to live. The idea of the arbitrariness of time had been yet understood by Saint Augustine as seen in these striking words (S. Aurelii Augustini, *Confessiones*, L.XI.C.XXIII):[47]

> Audivi a quodam homine docto, quod solis et lunae ac siderum motus ipsa sint tempora, et non adnui. Cur enim non potius omnium corporum motus sint tempora? An vero, si cessarent caeli lumina et moveretur rota figuli, non esset tempus, quo metiremur eos gyros, et diceremus aut aequalibus morulis agi, aut si alias tardius, alias velocius moveretur, alios magis diuturnos esse, alios minus? ... Sunt sidera et luminaria caeli in signis et in temporibus et in diebus et in annis. Sunt vero; sed nec ego dixerim circuitum illius ligneolae rotae diem esse, nec tamen ideo tempus non esse ille dixerit. ... Nemo ergo mihi dicat cae-lestium corporum motus esse tempora, quia et cuiusdam voto cum sol stetisset, ut victoriosum proelium perageret, sol stabat, sed tempus ibat.

We certainly are not concerned here, as elsewhere in this work, regarding the reasonableness of sidereal time and of the advantage of this time, as a consequence of the reasonableness of its introduction. These issues are outside the scope of the methods used in this work.

Having chosen the physical local time by a given method (having selected the basic movement), it will be no longer difficult to define any motion as a change in time of the coordinates of a moving point.

---

[47]I once heard a learned man say that the motions of the sun, moon, and stars constituted time; and I did not agree. For why should not the motions of all bodies constitute time? What if the lights of heaven should cease, and a potter's wheel still turn round: would there be no time by which we might measure those rotations and say either that it turned at equal intervals, or, if it moved now more slowly and now more quickly, that some rotations were longer and others shorter? ... Both the stars and the lights of heaven are "for signs and seasons, and for days and years." This is doubtless the case, but just as I should not say that the circuit of that wooden wheel was a day, neither would that learned man say that there was, therefore, no time. ... Let no man tell me, therefore, that the motions of the heavenly bodies constitute time. For when the sun stood still at the prayer of a certain man in order that he might gain his victory in battle, the sun stood still but time went on. Augustine, *Confessions*, Newly translated and edited by A. C. Outler, Book 11, Chapter XXIII, 29-30; this book is in the public domain: http://www.ling.upenn.edu/courses/hum100/augustinconf.pdf

It is quite clear that a transformation from one time to another will correspond to the replacement of the number t, which is the first time, by $\bar{t}$, which is the second time, and this replacement will be, of course, done by different methods at different points of space. In other words, the formula expressing the replacement of t by $\bar{t}$ should be written as

$$\bar{t} = f(x_1, x_2, x_3; t). \tag{13}$$

This formula gives a first indication of the more significant similarity of time and space than is commonly believed. Obviously, the physical local time at a given point is arithmeterized by a manifold of numbers; thus the physical time at a given point can serve as an interpretation of one-dimensional space. Since the choice of physical local time at a given point is arbitrary, which corresponds to the arbitrariness of the arithmetization of "one-dimensional space," whose interpretation is time, then, of course, the intrinsic properties of (physical) time should be, at a given point of space, invariant under the transformation given by formula (13). One of such intrinsic properties of the physical time is *the interval between two moments in time* – a notion which is analogous (for the one-dimensional "space" of time) to the notion of physical distance in three-dimensional physical space.

5. The interval between two moments of time is always implying that we are dealing with the same point in space and with different values of the physical local time, corresponding to different moments. The interval between two moments of time is a measurable intensity. The initial value of the time interval is the interval between two coinciding moments. The unit of measurement is chosen in a certain way in relation to the speed of light. The interval between two moments, differing from each other by an infinitesimal duration of physical local time, is called *infinitesimal interval*. Since we will need in the future only intervals between two infinitely close moments, we will consider only their definition.

Figure 16

To introduce the notion of interval between two infinitely close moments of time we will describe first of all a special instrument, the so-called *light clock*.[48] A light clock consists (Fig. 16) of a light

---

[48]The light clock is an ideal instrument, which differs completely from what we

source S, sending a light beam in a certain direction; a mirror m perpendicular to the beam (which reflects the beam in a direction, that is opposite to the direction of the incident beam), which could be moved at any distance $\tau/2$ from S, and, finally, a receiver S', located at the same point as S, which records the arrival of the reflected beam at S'. We will call the length, that is twice the distance from the light source to the mirror, *length of the light beam.*

We will call infinitesimal interval between two moments of time at a given point M in space such an infinitesimal length $\mathrm{d}\tau$ of the light beam in the light clock, in which the source S emits a signal at the first moment, and the receiver S' detects the reflected signal at the second moment. It is obvious that at different points M in space $\mathrm{d}\tau$ will be different, because, generally speaking, the speed of light will be different at different locations. If, while measuring the intensity of the interval, we choose another unit of measurement, then the quantity that represents the interval will not be $\mathrm{d}\tau$ but $l\,\mathrm{d}\tau$, which is obviously not essential. Comparing the infinitesimal interval $\mathrm{d}\tau$ with the infinitesimal increment of the local physical time $\mathrm{d}t$, we will have

$$\mathrm{d}\tau = T(x_1, x_2, x_3; t)\,\mathrm{d}t, \qquad (14)$$

where T depends, of course, both on the position of the point M (i.e., on the coordinates x_1, x_2, x_3) and on the moment t. The function T will be determined experimentally by using our light clocks at different points in space and at different times. The interval, of course, does not depend on the method of representing the physical local time. If we fix the mirror m in the light clocks so that the length of the light beam is equal to $\mathrm{d}\tau$, then no matter how we would introduce the physical local time, the interval between the emission and detection of the light signal will be (by definition) equal to $\mathrm{d}\tau$. This circumstance can be expressed as a requirement of invariance of the speed of light under the transformations (13). Since we related the interval of time with the motion of the light, we thus attributed a special role to light; this is not surprising, since in the definition of physical length we would ascribe a particular role either of light or of instruments (solid bodies), which serve to measure the physical distance (this issue is elementary and we will not discuss it here). The convenience and the important property of light clocks is that they can be used anywhere in the material space, because light (electromagnetic waves and current) moves (propagates) in all material bodies. Sound clocks cannot serve as clocks since they

---

called above a clock at a given point. A clock at a given point serves only as an auxiliary device, whereas light clock will, on the contrary, play a significant role related to the special role played by light.

could not be used, for example, in those parts of the material space, which are filled only with radiant energy. They could not be used to determine intervals of time in interplanetary space or in the space between molecules, and thus, the notion of interval of time determined by sound clocks would altogether be absent in many points of space. Such a quite natural generalization can be easily obtained when we consider closer the relationship of physical space and physical time – a relationship, expressing movement. Motion of any material

§6 Motion

1. By considering the physical space and time separately, we have seen how the arbitrariness of the arithmetization of both space and time led us to recognize that the intrinsic properties of space and time, i.e., the properties that do not depend on the method of arithmetization, should be invariant with respect to the following transformations of the numbers x_1, x_2, x_3 and t, arithmetizing space and time by one method, into the numbers $\overline{x}_1, \overline{x}_2, \overline{x}_3$, and $\overline{t}$, arithmetizing space and time by another method

$$
\begin{aligned}
\overline{x}_1 &= f_1\left(x_1, x_2, x_3\right), \\
\overline{x}_2 &= f_2\left(x_1, x_2, x_3\right), \\
\overline{x}_3 &= f_3\left(x_1, x_2, x_3\right), \\
\overline{t} &= f\left(x_1, x_2, x_3; t\right).
\end{aligned}
\tag{15}
$$

It is easy to see that these transformations show the separate role of time and are not the most common transformations of one quadruple of numbers, arithmetizing space and time, into another quadruple of numbers .

Already the very unnatural form of the formulas (15) shows the need for their further generalization. Such a natural generalization can be easily obtained when we look closer at the link of physical space with physical time – a link manifesting itself through motion. The Motion of any material point can be expressed analytically by the fact that the coordinates of this point are functions of physical time

$$
x_1 = \varphi_1(t) \quad x_2 = \varphi_2(t) \quad x_3 = \varphi_3(t).
$$

The properties of motion are studied in a special branch of mechanics – *kinematics*, which is justly called *geometry of motion* and which becomes real geometry, if we interpret the three coordinates x_1, x_2, x_3 and time t, as the four coordinates of a four-dimensional space: in this interpretation the motion of any material point is given in the form of a *curve* in this four-dimensional space; this curve completely determines the motion, i.e., the life of our material point, and therefore can be called and life curve of the material point. We will follow the adopted terminology and will call this curve *the worldline of our point.*

To imagine clearer this interpretation of motion, representing kinematics as geometry, let us simplify our space by regarding it as a two-dimensional Euclidean plane and arithmetize this space by introducing orthogonal rectilinear coordinates (x_1, x_2). To interpret motion, which takes place in our plane, we consider a three-dimensional Euclidean space and arithmetize it by introducing orthogonal rectilinear coordinates so that the two coordinates would be x_1, x_2, and the third would be time t (Fig. 17). The motion in our plane will be expressed by the equations:

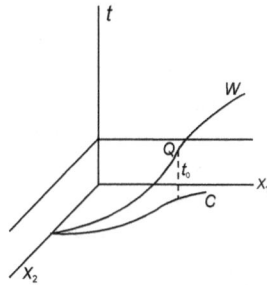

Figure 17

$$x_1 = \varphi_1(t) \quad x_2 = \varphi_2(t).$$

The geometric representation of the positions of a moving point gives us the trajectory C, but the trajectory alone does not completely determine the motion, because we cannot say by using only the trajectory at what of its points at a given moment of time our material point will be. On the contrary, the worldline W of a material point, which in our case is a curve in space (which has one time and two space dimensions), determines completely the motion of the

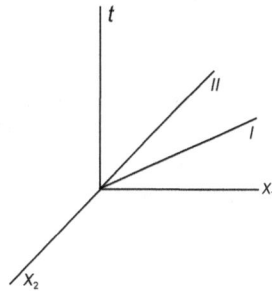

Figure 18

point. In fact, if we take a point Q on the worldline W and draw a perpendicular to the plane (x_1, x_2), we will find a point on the trajectory, where our material point will be at the moment t, which is determined by the length of this perpendicular. The plane $t = t_0$ will be a plane parallel to our plane (x_1, x_2); the intersection of the plane $t = t_0$ with the worldline W determines the position of our material point on the plane. Thus, the motion of all points of our plane will be completely determined, if we intersect the system of worldlines of our points by planes $t = t_0$ and assign different values to t_0.

Assume (Fig. 18) that two pedestrians move along the axis x_1, one at a constant velocity a and the other at a constant velocity b; both start from the origin at moment $t = 0$. The trajectories of both pedestrians will be the same, but their worldlines (lines I and II) will be different; it is also easy to see (Fig. 19) that the worldline of a point moving uniformly in a circle, is a helix. All that has been said about a particular method of arithmetization of two-dimensional space and time, is also applicable to the general case of an arbitrary arithmetization of space and time.

By specifying the worldline of a material point in a four-dimensional space, we will know the four coordinates, i.e., the numbers x_1, x_2, x_3 and t of each point of this world-line, and will therefore know the position of our point at the moment t. We can determine the positions of all points of the three-dimensional world at the moment $t = t_0$, if the system of worldlines corresponding to these points, is intersected by a three-dimensional space, which

Figure 19

is obtained from the four-dimensional space by setting $t = t_0$; such a space is a special case of what we will call, from now on, a *hypersurface* of the space of four dimensions. Thus, the entire life of our physical space (i.e., the whole durations of the motions of the material objects in it) can be expressed by a system of worldlines in a four-dimensional space.

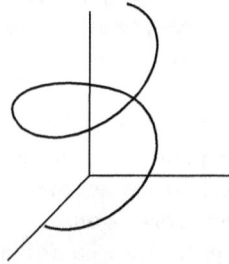

2. Speaking about motion, we considered motion in physical space, i.e., in the space filled with material objects, in which the position of a point is determined by material objects or processes; so the motion, we consider, is a *relative motion* with respect to those material objects which serve as the "coordinate grid" for our physical space. However, it is obvious that the physical space could be arithmetized by using not bodies which are stationary, but by moving bodies; then the second class of bodies would form a physical space with respect to which the bodies, which were first regarded as stationary, would move. It would be useless to try to answer the question of which bodies move and which are stationary, because the very term "move" can have only relative meaning: it is only possible to consider the motion of a system of bodies in relation to another system. To try to answer the question of what is the true motion, is equivalent to try to answer the meaningless question of what to what is attached: the tail to the dog or the dog to the tail.

The question now is, in what way are the points of our three-dimensional world moving? For the representation of this motion, through each point (x_1, x_2, x_3) of the three-dimensional space it is necessary to draw a worldline; this worldline will represent the motion of a material point; a given point on the worldline represents our material point at time t and space coordinates

$$\overline{x}_1 = \varphi_1(t), \quad \overline{x}_2 = \varphi_2(t), \quad \overline{x}_3 = \varphi_3(t).$$

Since to each point of the three-dimensional space corresponds a world-line, then, obviously, the form of the functions $\varphi_1, \varphi_2, \varphi_3$ will be different for the different points of space, and, therefore, the functions will depend not only on t but also on (x_1, x_2, x_3)

$$\overline{x}_1 = f_1(x_1, x_2, x_3; t), \quad \overline{x}_2 = f_2(x_1, x_2, x_3; t)$$

$$\overline{x}_3 = f_3(x_1, x_2, x_3; t)$$

One of the methods of arithmetization of space is to assign the numbers (x_1, x_2, x_3) to each of its points. In this method, material points move relative to the points of our space according to the above formulas, but we might regard the material points as stationary and then we could arithmetized space by assigning three new numbers $(\overline{x}_1, \overline{x}_2, \overline{x}_3)$ to each of its points. Thus, space and time can be arithmetized by two methods – either with the quadruple of numbers (x_1, x_2, x_3) and t or with another quadruple of numbers $(\overline{x}_1, \overline{x}_2, \overline{x}_3)$ and $\overline{t}$. The transition from one method of arithmetization to another is expressed by the formulas:

$$
\begin{aligned}
\overline{x}_1 &= f_1(x_1, x_2, x_3; t), \\
\overline{x}_2 &= f_2(x_1, x_2, x_3; t), \\
\overline{x}_3 &= f_3(x_1, x_2, x_3; t), \\
\overline{t} &= f(x_1, x_2, x_3; t).
\end{aligned}
\tag{16}
$$

All intrinsic properties of space and time must have invariant forms with respect to the transformation (16); in these transformations the role of time does not particularly differ from that of the space coordinates and it is sufficient to denote the time t by x_4, and the function f by f_4 in order that even the very form of the transformations (16) would no longer refer to the special role of time. In terms of these notations, we can rewrite (16) as follows:

$$
\begin{aligned}
\overline{x}_1 &= f_1(x_1, x_2, x_3, x_4), \\
\overline{x}_2 &= f_2(x_1, x_2, x_3, x_4), \\
\overline{x}_3 &= f_3(x_1, x_2, x_3, x_4), \\
\overline{x}_4 &= f_4(x_1, x_2, x_3, x_4).
\end{aligned}
\tag{17}
$$

Thus, time is overthrown from its pedestal. This is a fulfillment of the words of the great German mathematician Minkowski, used as epigraph to this chapter, and the physical world appears before us in its true light, as a collection of things called *events*, which in the arithmetization of this four-dimensional world, are characterized by the four numbers (x_1, x_2, x_3) and t. This means that the *physical world* can serve as the interpretation of the space of four dimensions; an *event* of the physical world becomes the interpretation of a point of the four-dimensional geometrical space. This new point of view of the physical world makes it possible to overcome the difficulties of its studies, which we pointed out at the end of the previous chapter: time ceases to interfere with our studies, on the contrary, having lost its privileged position and considered on equal footing with the spate coordinates, time becomes an active assistant in the study *no longer of the physical space and no longer of the physical time, which by themselves do not exist, but of the manifold spacetime – the physical world.*

The intrinsic properties of the physical world are invariant under the transformations (17), and if we recognize that all processes of the physical world are governed by laws that do not depend on the method of arithmetization of the physical world, all *these physical laws will be intrinsic properties of the physical world and will have a form invariant with respect to transformations (17).* This fact, which we will call the postulate of invariance, plays a huge role in the study and establishment of physical laws.

3. It is useful to look, at least briefly, at the historic path that led to the concept of the four-dimensionality of the physical world and to the postulate of invariance. This historical path, which is the thorny path of discovering the truth, was quite different from the logical path which we followed above.

The question of arbitrariness of the physical time was not initially raised at all; it was taken for granted that there was one time – the universal time, which the stars showed us. The situation was different with the coordinate system for the space; for many centuries the privileged position of the Earth seemed to be unshakable and lasting, for centuries the dominated view was the Ptolemaic picture of the world. Copernicus, one of the first who pushed the Earth in space and made it move, transferred the coordinate system to the Sun, and proposed to study the motion of the physical world with respect to this system. Galileo and Newton, who established the principles of mechanics, formulated and clarified the well-known and felt by all the law of inertia. The true meaning of this law is that by using the laws of mechanics we cannot detect uniform and rectilinear motion; in two physical spaces, one of which is stationary with respect to a third space, and the other moves uniformly and rectilinearly, all mechanical phenomena will be *exactly the same*. By choosing as the arithmetization of space a system of orthogonal rectilinear coordinates and assuming that the motion of one system relative to another is along the x_1−axis at a constant velocity v, we can express the law of inertia as the requirement of invariance of the laws of mechanics with respect to the transformations of the form:

$$\overline{x}_1 = x_1 - v\,t,$$
$$\overline{x}_2 = x_2,$$
$$\overline{x}_3 = x_3, \tag{18}$$
$$\overline{t} = t.$$

The development of our knowledge of the physical world added to the laws describing mechanical phenomena also the laws of electromagnetic phenomena, including that of the propagation of light which is a special form of electromagnetic waves. It was found out experimentally that space (ether), where electromagnetic (light) waves propagate, does not move with the Earth (Fizeau's experiment, etc.); thus, although it is impossible from a closed room on the Earth's surface to detect, by mechanical phenomena, the almost rectilinear and uniform motion of the Earth along its orbit (for small segments of it), but it seemed that that would not be difficult to achieve with the help of light phenomena, since light propagates in space, which does not move together with the Earth; thus by using light it would be as easy to detect the motion of the Earth as the motion of a boat (no matter how uniform it may be) with respect to stationary riverbanks. In 1881 Michelson performed his famous experiment, intended to detect the motion of the Earth relative to the space in which light propagates. The result of Michelson's experiment was negative: it failed to notice any effect of the Earth's motion on the light waves which propagate in a laboratory on the Earth's surface and therefore moving in space together with Earth. It was necessary to assume that the velocity of light is the same no matter whether we consider the propagation of light relative to a stationary space or relative to a space moving uniformly together with the Earth. This hypothesis, in connection with the law of inertia, makes it possible to correct formulas (18), expressing the transformations of coordinates in rectilinear uniform motion, by adding to these formulas some small quantities, which, in ordinary cases, almost do not affect the result numerically, but have tremendous fundamental importance. This correction to the transformation formulas (18) first placed time on a par with the space coordinates, and it was namely this very small correction which led to the end of the reign of time. With the correction, the transformation (18) can be written as

$$\overline{x}_1 = \frac{x_1 - v\,t}{\sqrt{1 - \frac{v^2}{c^2}}},$$

$$\overline{x}_2 = x_2,$$

$$\overline{x}_3 = x_3, \tag{19}$$

$$\overline{t} = \frac{t - \frac{v}{c^2}\,x_1}{\sqrt{1 - \frac{v^2}{c^2}}},$$

where c is the speed of light and the quantity v/c is negligible at our usual small speeds, so formulas (19) are practically almost indistinguishable from formulas (18). Essentially, however, the difference is huge, because the last of the formulas (19) has the character of the formula (13) and immediately introduces the concept of local time; in addition, in formulas (19) time already ceases to play the exceptional role as in formulas (18). Formulas (19) laid the foundation of the so-called special principle of relativity, introduced by Einstein and named so to reflect the assertion of the theory, based on this principle, that it is impossible to determine uniform linear motion with respect to space.[49] The special principle of relativity for the first time began to consider space and time not separately, but unified in the form of the physical world, whose laws must have a form invariant with respect to transformations (19).

The special relativity principle gave a number of consequences, whose interpretation in the physical world has undergone experimental verification and brilliantly passed that test.

In connection with the special principle of relativity, which rejected any possibility to detect uniform motion in space by physical experiment, an idea could occur, and it did actually occur to Einstein, that it is impossible to discover the uniform motion of a system from inside the system itself. Suppose that a system A moves relative to a system B. If A moves non-uniformly and rectilinearly or along a curved path or with any acceleration, then in system A will be observed phenomena which are not observed in the system B; for example, the initial

---

[49]EDITOR'S NOTE: Expressed differently, the assertion of Einstein's theory is that absolute motion (motion with respect to space) does not exist, and the only motion manifested in phenomena is relative motion (relative to physical objects).

acceleration of an elevator is perfectly felt by an observer in the elevator; observations of a swinging pendulum (Foucault's experiment and the Pantheon) show the rotation of the Earth, all of whose points, describing a circular trajectory, move, of course, with an acceleration (centripetal). Assume that, by experimenting, we have found a number of phenomena in the system A, occurring as if the system A moved with an acceleration with respect to the system B. Can we conclude from this that the system A moves with respect to the system B, and not vice versa? Does this statement make any sense at all? If the forces acting on bodies in system A, which arise as a result of its apparent motion, and such forces are not present in B, then we can certainly conclude that A is accelerating with respect to B. But we can alternatively assume that the forces acting in the system A exist everywhere; they would act in B, if B did not accelerate relative to A, but the accelerated motion of B would destroy the forces acting everywhere. To have a clearer idea of our difficulty, let us imagine that we, using very accurate devices, would notice on the Earth the so-called deviating force arising from the Earth's rotation, or the Coriolis force acting on all bodies which are in motion on the Earth's surface. If the Earth was always covered with a layer of clouds, which were preventing light from stars, planets, the Moon and the Sun from reaching the Earth's surface, we can unmistakably say that the very idea – to regard the Coriolis force as an effect of the Earth's rotation – would seem to be absurd and useless; this Coriolis force would be a property of moving material bodies, which resembles the property of universal gravitation. In systems where there would be no Coriolis force, we would explain this absence as a result of the motion of this system with respect to our system on Earth, from where we cannot see the light from the sky. This shows that not only can we not detect the uniform and rectilinear motion of the system, sitting inside it, but we cannot determine either which of two systems in relative motion is moving and which is stationary. It is impossible to decide who is right[50] – Ptolemy or Copernicus; it is impossible, of course, if we do not resort to the principles of reasonableness, economy of thought, etc. One of the cleverest evidence of the correctness of the Copernican system is described in

---

[50] EDITOR'S NOTE: As indicated in the Editor's note on p. 40, Minkowski stressed in his talk "Space and Time" that in the four-dimensional world there exists an objective distinction between uniform and accelerated motion – the worldline of a uniformly moving body is straight, whereas the worldline of an accelerating body is curved. This sharp geometrical distinction reflects the experimental fact that accelerated motion of a system can be detected inside the system, whereas uniform motion cannot be discovered in any way.

52

the following poem by Lomonossov:[51]

On the motion of the Earth
(Epigram on opponents of Copernican system)

Two astronomers found themselves together
on the feast and extremely disputed in fervor.
One keeps saying "The Earth, while turning, goes around
the Sun";
The other, (is saying), that the Sun leads all planets with
itself.
First was Copernicus, other said to be Ptolomy.
Then a cook decided the dispute with a grin.
The master asked: "You know the stream of stars?
Say, how you are thinking about these doubts?"
He gave such an answer: "I prove it true, even I never was
on the Sun,
that Copernicus is right in his argument,
Who has seen such simpleton of cooks,
Who turned the hearth around the roast?"

The principle of reasonableness can clearly be recognized in these witty words.

So, the laws of the physical world must be invariant with respect to the coordinate transformation corresponding to any motion; hence there is only one step to the postulate of invariance. This step can be made, as we saw above, through a careful consideration of the idea of physical time. Thus, the longer historical path led us also to the four-dimensional physical world and to the postulate of invariance.

4. *The physical space and the physical time merged into the physical world, which interprets the geometrical space of four dimensions. The laws of the physical world are its intrinsic properties and are governed by the postulate of invariance.* Time does not differ in any way from the other coordinates.

Is the last assertion above correct? A more careful consideration will show that the last assertion cannot be considered entirely correct, and that the postulate of invariance, as well as the method of arithmetization of the physical world, should be restricted in a certain way in order to restore the distinct role of time. I cannot elaborate on this issue any further, which is not too developed in the theory of relativity. I will only note that the reason for the return of time to its distinct

[51]EDITOR'S NOTE: Translated by Sergey Andronenko, St. Petersburg, Russia.

role is *the principle of causality*, according to one of its requirements it is impossible, by changing the arithmetization of the physical world, to interchange cause and effect. This principle (formulated, of course, clearly and rigorously, and not as it was now done) should impose certain restrictions:

1) on the methods of arithmetization of the physical world, in which the postulate of invariance holds; any arithmetization can, of course, be used, but the postulate of invariance does not hold for any arithmetization;

2) on the properties of the geometrical four-dimensional space, whose interpretation is the physical world;

3) on the choice of that coordinate of the physical world, which will be regarded (in accordance with the principle of causality) as time.

Without discussing any further the apparently conserving features of time and clarifying in sufficient detail the notions of physical space, time, motion and the world, we arrive at their geometrical, so to speak, interpretation.

§7 The World

1. Considered from a purely abstract point of view, *time* is a collection of things called moments, which are in certain relations with one another and with the (three-dimensional) space, which relations may be established by special kinematical axioms and their consequences. Considered by itself, time constitutes what could be called "space" of one dimension; considered in connection with the three-dimensional space, time forms a space of four dimensions. (Abstract) time can be arithmetized completely arbitrarily; a certain number t will always be assigned to every moment. The transition from one arithmetization of time, considered by itself, to another can be expressed by the replacement of the number t with $\bar{t} = f(t)$. Abstract time as "space" of one dimension will have its metric. Let us call the infinitesimal "distance" between two moments infinitesimal interval of time and denote it by $d\tau$. If, when arithmetizing time, the specified moments are represented by the numbers t and $t + \Delta t$, then $d\tau = T\,dt$, where T depends on t; needless to say that the interval of time $d\tau$ does not depend on the method of arithmetization of time. If, however, we consider the abstract time not by itself, but in connection with the three-dimensional space, then the resulting spacetime can be again arbitrarily arithmetized by using quadruples of numbers; then to a given point of the three-dimensional space and a given moment of time will correspond

one and only one quadruple of numbers: x_1, x_2, x_3, x_4. The transition from one method of arithmetization to another will be expressed by the replacement of one quadruple of numbers x_1, x_2, x_3, x_4 with another quadruple: $\overline{x}_1, \overline{x}_2, \overline{x}_3, \overline{x}_4$. As one quadruple completely defines the other and vice versa, the we will have the following formulas of the transformation:

$$
\begin{aligned}
\overline{x}_1 &= f_1\,(x_1, x_2, x_3, x_4), \\
\overline{x}_2 &= f_2\,(x_1, x_2, x_3, x_4), \\
\overline{x}_3 &= f_3\,(x_1, x_2, x_3, x_4), \\
\overline{x}_4 &= f_4\,(x_1, x_2, x_3, x_4).
\end{aligned}
\tag{20}
$$

Let us call the four-dimensional space, about which we just talked[52] (spacetime), *geometrical world* or simply *world*.

A certain point of the world, which we will call sometimes *event*, corresponds to a given point of space and a given moment of time. Everything that was said in the previous chapter can be transferred to the four-dimensional space, and thus we can define the geometry of this four-dimensional space, in other words, the *geometry of the world*. The non-intrinsic properties of the world will depend on the method of its arithmetization, whereas the intrinsic properties of the world (e.g., the magnitude of the distance between two points, the vectorial and metric curvature) will not depend on the way the world is arithmetized and will be, therefore, expressed in a form which is *invariant* (in the sense discussed above) with respect to the coordinate transformations (20). Let us call *interval* the distance between two points (events) of our world, the distance between two infinitely close points of our world will be an *infinitesimal interval*; the metric of the world (and hence the definition of the infinitesimal interval) will depend on ten quantities – the components of the fundamental metric tensor. Parallel transport in our world may depend, in addition to the ten quantities, on the four components of the scale vector. It should be noted that the term "interval" should not yet be given any specific content.

2. Let us choose, for the moment completely arbitrarily, one of the world coordinates as the *time coordinate*, and enumerate the coordinates in such a way that the time coordinate is always denoted by x_4. We will call the other coordinates x_1, x_2, x_3 *space coordinates*. In our world, we distinguish, like in three-dimensional space, two geometric loci of points; one of these loci consists of points satisfying the relation

$$
f\,(x_1, x_2, x_3, x_4) = 0.
$$

---

[52]In contrast to the analyzed above physical world.

We will call this geometric locus (analogous to a surface in our space) *hyper-surface of the world.*

Of course, it is not difficult to note that a hyper-surface of the world will possess the properties of a space of three dimensions, but the term space will be retained for other entities and will not be used for hyper-surfaces at all. Of all hyper-surfaces, of special interest for us is the hyper-surface whose equation has the form $x_4 = x_{40}$ and which is obtained by setting the time coordinate equal to constant. This hyper-surface will be a geometric locus of events of the world, corresponding to a given value of the time coordinate. The points of this hyper-surface will have arbitrary space coordinates, and we will call this hyper-surface the *space* corresponding to the given moment x_{40}. Another geometric locus of points of our world will be analogous to a curve of our space and will be called a *world curve.* The world curve will be a geometric locus of points, defined by the equations

$$x_1 = \varphi_1(u), \quad x_2 = \varphi_2(u), \quad x_3 = \varphi_3(u), \quad x_4 = \varphi_4(u),$$

where u is an arbitrary parameter. By excluding u from the above equations we will obtain three relations between the coordinates x_1, x_2, x_3, x_4 of the world. The simplest world curve is the curve, in which the space coordinates in these three relations are assigned constant values

$$x_1 = x_{10}, \quad x_2 = x_{20}, \quad x_3 = x_{30}.$$

The different points of this curve will correspond to different values of the time coordinate; such a world curve will be called *time which corresponds of a given point* (x_{10}, x_{20}, x_{30}) *of space.* The world curves are naturally divided into two classes. Those world curves for which the time coordinate is constant $x_4 = x_{40}$ (does not depend on the parameter u), are called *space lines or trajectories.* A space line lies entirely in space, corresponding to a certain moment (namely on the hyper-surface $x_4 = x_{40}$), and is a typical example of the mentioned above curves in the space of three dimensions. The world curves, which do not belong to the space lines, form another class of curves, whose time coordinate x_4, depends on the parameter u. Therefore u can be expresses as a function of x_4 and the equations of the world curves, excluding u, give the following relations

$$x_1 = \psi_1(x_4), \quad x_2 = \psi_2(x_4), \quad x_3 = \psi_3(x_4).$$

These relations, which coincide completely (if x_4 is replaced by t) with the equations, which in the previous section served for defining world lines in the physical world. The world curves of the second class will be called *time worldlines* or simply *worldlines*. We will call *motion* a phenomenon, which is completely determined by a worldline; thus the expression "given motion" will be completely equivalent to the expression "given worldline."

It is easy to see, that all our definitions are so chosen, that a term assigned to an object of the physical world, serves as an interpretation of the notion denoted by the same term in the geometrical world. For example, motion in the physical world is an interpretation of motion in the geometrical world; a worldline in the physical world interprets a worldline in the geometrical world. The geometrical space does not have an independent significance, being only a hyper-surface of the geometrical world, which completely corresponds to the fact that space does not exist as an independent (separate) entity since it is unthinkable without the physical time.

3. Let us now turn to the metric of the world, i.e., to the definition of the infinitesimal interval, which for brevity is denoted by $d\sigma$; the square of the infinitesimal interval depends on the squares and the products of dx_1, dx_2, dx_3, dx_4 and also on the fundamental metric tensor of the world. We can write the expression for the interval

$$
\begin{aligned}
d\sigma^2 = {} & g_{11}\, dx_1^2 + g_{22}\, dx_2^2 + g_{33}\, dx_3^2 \\
& + 2\, g_{23}\, dx_2\, dx_3 + 2\, g_{31}\, dx_3\, dx_1 + 2\, g_{12}\, dx_1\, dx_2 \qquad (21) \\
& + 2\, g_{14}\, dx_1\, dx_4 + 2\, g_{24}\, dx_2\, dx_4 + 2\, g_{34}\, dx_3\, dx_4 + g_{44}\, dx_4^2.
\end{aligned}
$$

The way the square of the interval is written demonstrates the distinct role of the time coordinate. We can easily see that in two special cases our interval becomes either something resembling distance ds or something resembling the duration between two moments $d\tau$. Consider the infinitesimal interval $d\sigma_1$ between two events, having *the same* time coordinate i.e., lying in a certain (the same) space, in the sense discussed above. For these events x_4 will be constant, thus, dx_4 will be equal to zero and the formula for $d\sigma^2$ will become formula (9) for the square of infinitesimal distance ds^2: $d\sigma_1^2 = ds^2$ or $d\sigma_1 = ds$ (since we can always take care of the sign). Since we regard space only as a hyper-surface of the word, we can also determine the metric of space from $ds = d\sigma_1$, which results from $d\sigma$ by setting $dx_4 = 0$.

Consider now the infinitesimal $d\sigma_2$ between two events, having the same space coordinates, i.e., such events whose time coordinates are different and for which therefore $dx_1 = dx_2 = dx_3 = 0$. The interval between such events is defined by the formula $d\sigma_2^2 = g_{44}\, dx_4^2$. As we regard time only as a certain worldline we can determine the metric of time from the duration $d\tau$ whose square can be obtained by setting $dx_1 = dx_2 = dx_3 = 0$ and *changing the sign in the resulting expression.* In other words, for $d\tau^2$ we have $d\tau^2 = d\sigma^2 = -g_{44}\,dx_4^2$, $d\tau = \sqrt{-g_{44}dx_4}$. Comparing this formula with expression (14) for $d\tau$ in §5, we find that $T(x_1, x_2, x_3, x_4) = \sqrt{-g_{44}}$ where in (14) t is replaced by x_4.

It is clear from the above that the metric of the world itself also defines the metric of space for any moment of time and the metric of time for any point of space. The metric of the world is not completely arbitrary; we require that the distances in space be expressed by real (not imaginary) numbers, and in the same way a period of time should be also expressed by a real number. This requirement can be summarized in the following *postulate of realness* of space and time.[53]

The world metric should be such that 1) the interval would be real at all points of the world and for all events, having the same time coordinate (the square of the interval in this case should be positive), and that 2) the interval should be purely imaginary at all points of the world and for all events, having the same space coordinates (the square of the interval in this case should be negative).

The postulate of realness of space and time has extremely strong influence on the metric of the world; this postulate is intrinsically linked to the principle of causality. This postulate allows to divide the direction of the world curves into two classes: *space-like* and *time-like* directions; the first are characterized by the fact that the magnitude of the infinitesimal four-dimensional vector characterizing them (i.e. of the corresponding interval) is positive, whereas for the second, this magnitude is negative. Every space line has space-like direction, whereas every world curve, which we called time, has time-like direction. Space-like directions are separated from time-like directions by a special hyper-surface (something like a cone); on this hyper-surface, called *null-cone*, the intervals between its points are all zero. The transition through the null-cone corresponds to the transition from space to time. Not being able to discusses these issues in more detail, I will note that the notion "angle," which has a completely clear meaning for two space-like or two time-like directions, should be modified when considering the angle between a space-like and a time-like direction. In short, the null-cone forms as if a boundary in the world, which separates one part of the world from another, having very spe-

---

[53] I have simplified the postulate a bit to make it more obvious; in fact, it should be supplemented by the conditions following from the application of the so-called "law of inertia" to the form for $d\sigma^2$. See, for example: D. Hilbert, *Die Grundlagen der Physik*, Zweite Mitteilung (1917).

58

cial properties.[54]

The postulate of realness imposes certain conditions on the arithmetization of the world: the arithmetization should be such that in the new coordinate system it can be also possible to select space and time coordinates, and in the new coordinates the postulate of realness should hold; we will call such arithmetization *selected*.

With the help of the postulate of realness it can be proved that two events lying on a worldline, corresponding to different moments of time, cannot, by changing the arithmetization, become simultaneous as long as the arithmetization, as it changes, remains selected; thus, two events that could be regarded as cause and effect, can never be made, by changing the arithmetization, simultaneously; in other words, these two events will always be to each other in relation of cause and effect.[55]

4. Now it remains for us to turn to the question of how, by studying experimentally the physical world, we can determine the metric of the geometrical world, whose interpretation is the physical world. First of all, we have to decide how the physical world interprets the geometrical one. Suppose that we created a dictionary with which we can figure out what object of the physical world interprets a given notion (thing) of the geometrical world. This dictionary completely depends on out will. Thus, we can interpret the geometrical world, one way or another, with the help of the physical world. But suppose that we have chosen a given method of interpretation and agree that the same term refers to both the notion (thing) of the geometric world and to the object of the physical world which interprets the notion.

Once this is done, then, by measuring intervals in the physical world, we will be able to investigate experimentally the metric of the geometrical world (see §3). Further, by parallel transport of a vector along closed lines in the physical world, we determine its vectorial and metric curvature, and thus determine the scale vector and the properties of parallel transport in the geometrical world (see §4). In principle, therefore, the problem of studying the geometrical world, through its

---

[54]The null-cone is some kind of singularity in space for multiple-valued functions $\sqrt{()}$.

[55]It should be noted that any two events A and B with different time coordinates are regarded as cause and effect, where A can be viewed either as a cause or an effect of B. If A is a cause of B, i.e., if $x_{4A} < x_{4B}$, then replacing x_4 with $\bar{x}_4 = -x_4$, which transformation gives us again selected arithmetization, we will have $\bar{x}_{4A} > \bar{x}_{4B}$, i.e., B will be the cause of A. Thus, by changing the sign of time, we can always change cause and effect. If the sequence cause-effect should remain unchanged, apart from the postulate of realness, other additional restrictions on the properties of the world and the arbitrariness of its arithmetization are necessary.

interpretation which is the physical world, is resolved. However, a question immediately arises: can we, people, carry out such experiments or in order to perform them we need to stop being people and become gods? The answer to this question can be comforting: we can do these experiments (not practical, of course, but in principle). In fact, while being an obstacle for us in the study of the physical space, time, broke even the idea of the separately existing physical space – that same time will serve us as a great helper when determining the interval where the time difference and the space differences participate equally.

However, while measuring the interval, we first of all encounter the fact that we cannot deal with time arbitrarily; so we can determine experimentally the metric the world only for a very limited time.

Further, the possibility of studying, in principle, the geometrical world encounters a significant defect of our experimental equipment, namely the difficulties in measuring the interval. We need somehow to get around both these difficulties, and to find those quantities whose measurement in the entire world would be achievable by our technical means and which at the same time could determine the properties of the geometrical world. These quantities were found by the brilliant idea of Riemann, developed by Einstein and Weyl, the idea of binding forces that determine the properties of the world; it was necessary to seek those methods for determining the geometry of the world in motions and in the forces arising and acting while in motion. The next chapter will be devoted to the analysis of this issue.

60

GRAVITATION AND MATTER

Es muss also entweder das dem Raume zu Grunde liegende Wirkliche eine discrete Mannigfaltigkeit bilden, oder der Grund der Massver-hältnisse ausserhalb, in darauf wirkenden bindenen Kräften, gesucht werden.

Riemann "Ueber die Hypothesen, welche der Geometrie zu Grunde liegen."[56]

§8. The Old and the New Mechanics

1. By studying motion, in other words, by considering worldlines, we can divide motion into two classes: one class referring to *inertial motion*, the other – to motion *caused by the action of forces*. The division of motion into these two classes is completely arbitrary[57] and depending on our will; in fact, we make this division, of course, by obeying the principle of reasonableness. Performing this division, we establish how we can arithmetize the force corresponding to any of the motions not by inertia. It is not difficult to see that the division of motion into these two classes reflects the essence of the law of inertia, or Newton's first law, whereas the arithmetization of the force is Newton's second law. We shall call the division of motion into two classes the

---

[56]B. Riemann, Über die Hypothesen, welche der Geometrie zu Grunde liegen. *Abhandlungen der Königlichen Gesellschaft der Wissenschaften in G′ottingen*, Berlin, 1843, 1892, S, 133–152. Translation: "Either therefore the reality which underlies space must form a discrete manifold, or we must seek the ground of its metric relations outside it, in binding forces which act upon it." B. Riemann, "On the hypotheses which lie at the foundation of geometry", in W. B. Ewald, *From Kant to Hilbert: A Source Book in the Foundations of Mathematics*, Volume II (Oxford University Press, Oxford 1996), pp. 652–661, p. 661

[57]EDITOR'S NOTE: This is not quite correct. As explained in the Editor's notes on pp. 40 and 51 the two classes of motion are clearly distinct in spacetime – inertial motion is represented by a straight worldline, whereas accelerated motion is represented by a curved worldline.

principle of inertia and see how it enters the old mechanics. Regarding the division of motion into two classes one recalls Aristotle's theory of "perfect motions" (circular motion had been considered perfect); Solomon had been right to exclaim: "What was, will be, and what was done, will be done – and there is nothing new under the Sun ..." (Ecclesiastes).

When the principle of inertia had been incorporated in the old mechanics it had been presupposed that space was Euclidean and that sidereal time had been regarded as the physical time. Uniform and rectilinear motion (with respect to the stars) belonged to the first class of motion, i.e. to motion by inertia. All other motions were called motions under the action of a force, where force was a notion corresponding to each noninertial motion of a material point and arithmetized in a certain way (using the acceleration).

Let us see how one can, with the help of the idea of worldlines, express Newton's principle of inertia. Consider again, for the sake of simplicity, a two-dimensional space – a plane, which is arithmetized by introducing orthogonal, rectilinear coordinates; then rectilinear and uniform motion is represented by the formulas: $x_1 = a_1 t + b_1$, $x_2 = a_2 t + b_2$, where a_1, a_2, b_1, b_2 are constants, and t is the sidereal time. Going to the world corresponding to our two-dimensional space, i.e. employing a three-dimensional Euclidean space, in which the axis t is perpendicular to the axes x_1, x_2 (see Fig. 17), we see that motion by inertia is represented by a straight worldline; if a worldline corresponding to a given motion is not straight, that motion is caused (by definition) by the action of forces.

The postulate of realness, discussed above, makes it impossible to give such a visualization of the world, analogous to the one just given. For us, however, the fact, that the old mechanics requires specific geometrical properties of the world, is sufficient; without discussing what those properties are, we will call the world with such geometrical properties Euclidean;[58] in this world, the principle of inertia of the old mechanics can be formulated in the following way:

In the Euclidean physical world only the particles, whose worldlines are straight lines, move by inertia. The motion of all other particles, whose worldlines are not straight lines, is caused by the action of forces.

It should be stressed that the notion "straight line" should be understood in the sense discussed above (see §4, which deals with parallel transport); in fact, it is easy to prove that these "straight lines" in the Euclidean world are the ordinary straight lines, whose equations in

---

[58]In this Euclidean world we have $\mathrm{d}s^2 = \mathrm{d}x_1^2 + \mathrm{d}x_2^2 + \mathrm{d}x_3^2 - \mathrm{d}x_4^2$ in a correspondingly chosen arithmetization of the world.

orthogonal, rectilinear coordinates are: $x_1 = a_1t + b_1$, $x_2 = a_2t + b_2$, $x_3 = a_3t + b_3$.

The principle of inertia of the old mechanics imposes certain restrictions on the physical world by requiring that it be Euclidean; abandoning this restriction is a specific feature of the new mechanics of Einstein, Hilbert, and Weyl.

2. In the new mechanics, the principle of inertia is defined in the same way as in the old mechanics; it is only necessary to abandon the requirement (in fact, arbitrary) for the world to be Euclidean. Thus, the new mechanics establishes the following principle of inertia:

Particles whose worldlines are straight worldlines of our world move by inertia. All other particles move under the action of forces.

What remains to be done in the new mechanics, like in the old mechanics, is to arithmetize the force. Unfortunately, we cannot go into sufficient detail on this question, as the detailed development of the concept of force in the new mechanics requires a fairly large mathematical apparatus. For our purposes, however, it is completely sufficient to remark that the arithmetization of force depends on two types of quantities: first of all this arithmetization depends on three numbers (they would be four), which characterize the degree of deviation of the worldline, representing the motion of a particle, from straightness; these numbers are called *accelerations*; in the case of the Euclidean world, these numbers reduce to the usual notion of acceleration.

But as the same motion can be performed by completely different material points, the second quantity in the arithmetization of force is a number, which distinguishes one material point from another and which is called *mass of the material point*. By using the number, representing the mass of a material point, we define the notion of mass of a material body; using further some geometrical properties of the body (volume) we define the notion density. The physical space, filled with all material bodies, has a certain density at each of its points at a given moment; thus, each point of the physical world is characterized by a special number ρ – density. Obviously, ρ can be different at different points of the world and therefore it is a function of the coordinates of the world points: $\rho = \theta(x_1, x_2, x_3, x_4)$.

By studying the worldlines of the material points in the physical world, we could, knowing the geometry of the world, to determine when the material points move by inertia and when they are subject to forces. Further, by examining these worldlines we can determine the forces acting on the material points. In this case it would be also easy to perform a sufficient number of experiments to figure out that one and only one force corresponds each worldpoint. Thus, the world would turn out to be a kind of *a force field*; of course, at a given point in space the forces would change both directly with time and due to the fact that different material points would pass through that point of space at different moments of time. But it is possible to establish that one and only one force correspond to each worldpoint (event); it is impossible for a given force to be simultaneously present and absent at a give point of space.[59] However, in order to determine the force field in the world, we need to establish the geometry of the world, but we saw that it comes with great difficulties. The impossibility to determine experimentally the geometry of the world at this moment compels us to make certain hypotheses about it. The old mechanics immediately makes an extremely narrow hypothesis that the world is Euclidean. By making this very strong constraint the old mechanics had won a lot; its laws acquired extremely simple character, and we owe the enormous advancement of our knowledge and technological culture to this simplicity. At first, the new mechanics tried to avoid making additional hypotheses about the geometry of the world. It could do it, but then would doom itself to a pity and futile existence for many centuries. To become fruitful, the new mechanics, due to the limitations of our experimental tools, like the old mechanics, needs additional hypotheses about the geometrical nature of the world. These hypotheses were made first by Einstein (the hypothesis of gravitation), and then, in a more general form, by Weyl (the hypothesis of matter). The nature of these hypotheses differs very much from the hypotheses of the old mechanics about the Euclidean world; the distinct feature of the new hypotheses, as expected from the progress of science, is their greater breadth and extent; they do not define the metric of the world right away as the old mechanics does, but indicate only certain properties of the metric, allowing experiment to determine the metric conclusively. We will now start studying these hypotheses.

§9 Gravitation

1. First of all, let us see what the old mechanics deduces from its hypothesis of the Euclidean nature of the geometry of the world. Consider, from the point of view of the old mechanics, the motion of a material point in the material space, which means in the presence of other material bodies. In some regions of space, a material

---

[59]To avoid misunderstanding, it is necessary to point out that a force acting at a given moment of time at a given point of space can be represented by several forces, but all these forces define one and only one resultant force at a given point of space at a given moment of time.

point will move almost by inertia, in others its motion by inertia will be disrupted and we will detect the action of forces. Special material objects, which we call gravitating masses (for example, the Sun, the planets and so on), cause very significant forces. By studying the deviation from the motion by inertia of our point in the presence of material objects, which we called gravitating masses, we can determine the action on our point of special forces induced by these gravitating masses. If our material point is characterized with the help of light rays, the action of the gravitating masses on them will be negligible and hardly measurable. But, if our material point itself belongs to the class of gravitating masses, the situation drastically changes and its motion will differ strongly from inertial motion. By studying this deviation from inertial motion we arrive at the law of universal gravitation. Namely, such an approach was followed by Newton while studying, with the help of Kepler's laws, the motion of material points representing two gravitating masses – the planet of Mars orbiting the Sun. In the usual terms of Euclidean geometry and the old mechanics the law of universal gravitation can be formulated in the following way: *on each of two gravitating masses acts a force of attraction, directed from one to the other gravitating masses and whose magnitude is proportional to the masses of the gravitating masses and inversely proportional to the square of the distance between them.*

If the material point moves not in the vicinity of gravitating masses, but near some other material entities, for example in a beam of electromagnetic waves (light), then such a point will not move by inertia either because it will be subject, for example, to the light pressure. In such a way, besides the forces of universal gravitation there may exist various other forces. If our point moves in the vicinity of material objects of electromagnetic nature, we will call such other forces *electromagnetic*. There may also exist some other forces, which we will call *additional*. Therefore, we reach the conclusion that three kinds of forces can act on a material point: 1) the force of universal gravitation, 2) electromagnetic forces, and 3) additional forces. It is the action of these forces that studied by mechanics and physics. Apparently, all forces acting on all kinds of material points (regardless of whether or not these points represent gravitating masses) are reduced either to forces of universal gravitation or to electromagnetic forces. Thus, it is self-evident that the arithmetization of our world id carried out in a given manner; for example we choose sidereal time, and the coordinate system is set by using the so-called immovable stars. One cannot talk at all about the principle of invariance in its general form in the old mechanics; once the coordinate system is chosen, the forces resulting from the Earth's rotation are, of course, fictitious forces and does not belong to the list of the discussed forces.

2. Let us consider in more detail the force of universal gravitation in the framework of the old mechanics. It turns out that by using this force we can study in great detail and precision and , if we can express it this way, to explain the motion of the planets of our system, and also a number of motions of systems of stars (binary stars and so on). Of course, the planets of our system, which are subject to the resultant influence of many gravitating masses and which are not points at all, move in a rather complex and entangled way than material points obeying the simple law of universal gravitation in the presence of a single gravitating mass. However, celestial mechanics perfectly coped with all the difficulties, and the vast majority of the planetary motions was explained by the law of universal gravitation.

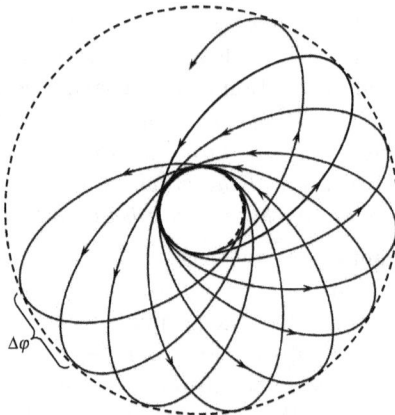

Figure 20

There existed, however, small details of the motion of planets that cannot be explained by the usual method of celestial mechanics. Such a detail is the unexplained part of the motion of Mercury's perihelion. The closest point to the Sun through which Mercury passes each year sifts (Fig. 20) and a small fraction of this shift (43″ in 100 years) has not been explained satisfactory by the classical celestial mechanics until recently; we should point out that of all planets Mercury is the closest to the Sun, and namely the motion of this planet was in the greatest disagreement with the principles of celestial mechanics. We indicated above that the old mechanics had to choose, in the arithmetization of the space of the world, a given coordinate system connected to the stars. For a possible fantastical explanation of the universal gravitation one could assume that the old mechanics could choose not the indicated coordinate system, but some other, relative to which the coordinate system connected with the stars, moves in a certain way; as a result of this motion some apparent forces arise, and namely these forces we observe in the phenomenon of universal gravitation. It is self-evident, that in the framework of the old mechanics this hypothesis would turn out to be indeed fantastical and it would be easy to show that no motion of our coordinate system, as a whole, could explain the phenomenon of universal gravitation. The situation would be significantly simpler, if universal gravitation degenerated into a constant force with a fixed direction, whose magnitude is equal to the mass times the *constant* acceleration g of the force of weight. It is such a degenerated force of universal gravitation (or rather "universal weight") that might be easier to explain kinematically. In fact, if our stellar coordinate system moved in a straight line and with uniform acceleration g relative to the coordinate system of the old mechanics, then in such a moving system we would observe an apparent force, which coincides with the force of universal gravitation. I will not discuss this in detail since this example of "universal weight" is described at length in all popular books on the principle of relativity. This example serves only to indicate the possibility to explain kinematically some forces, but it gives a completely wrong impression that the old mechanics can explain kinematically the force of gravity which is completely untrue.

For the kinematical explanation of the force of gravity a significant role, at least historically, played the experimentally confirmed with great precision fact, that the mass in the law of universal gravitation (gravitational mass), is identical with the mass in the Newton's second law (inertial mass), according to which the force is equal mass times acceleration. Eotvös' clever experiments showed that to within $\frac{1}{10^8}$ the gravitational and inertial masses are identical. If this fact is not accidental, it appears natural to try to explain the force of gravity kinematically and to regard this force as apparent (fictitious), or (by D'Alembert's principle) as inertial force. Inertial mass is a coefficient of proportionality in the inertial force, whereas gravitational mass is a coefficient of proportionality in the force of gravity, and since these masses are identical, it appears possible that the force of gravity is kinematical force. This hypothesis is impossible in the old mechanics and in the old views of the Euclidean nature of the world. Is it possible to use it in the new mechanics by making some kind of general assumptions about the geometry of the world, use these assumptions to link the metric of the world with gravitation, and in this way involve astronomy and celestial mechanics, with their amazingly precise methods, in the attempts to determine the geometry of the world experimentally? This is the way which leads to the foundations of Einstein's general principle of relativity.

3. We will now consider the hypotheses in Einstein's theory[60] about certain properties of the physical world and its geometry, which is interpreted as our physical world. These hypotheses can be formulated as three assertions:

1) *The Geometry of the world is Riemannian geometry.* This means that a straight line in the world is at the same time the shortest line, there is no scale vector, and the metrical curvature of the world is zero.

2) *Under the influence of gravitating masses a material point moves by inertia.*

In other words, the worldline of a material point, which is subject to the influence of gravitating masses, is a straight line in our world. This second hypothesis is a hypothesis as long as we found a process in the physical world which corresponds to parallel transport in the geometrical world. If this process is determined, then the straight line in our physical world is also determined. But we can regard this hypothesis as defining the process which corresponds to parallel transport in the geometrical world. Assume that we define the straight line in the physical world as the worldline of a point which is under the influence of gravitating masses, assume also that we can determine the physical directions, vectors and angles between them, assume that the vector a should be parallel transported from point M_1 to point M along the curve C in the physical world (Fig. 21). In order to carry out the parallel transport, we divide the curve C into infinitesimal parts by using points and connect neighboring pairs of points with straight lines (i.e. with the worldlines of points subject to the corresponding gravitational masses). Our vector a can be easily parallel transported along any of these straight lines, if we keep the angle between it and the straight line constant (see §4). By carrying out the parallel transport along each straight line we will bring our vector to point M, and this process will be as close to parallel transport along a curve as the number of parts into which the curve is divided is greater. In this way,

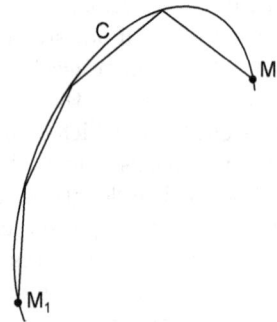

Figure 21

---

[60]Einstein does not pay much attention to the logical aspect of the theory – in his works these hypotheses are not properly formulated.

we can really regard the second hypothesis as establishing a process in the physical world, which corresponds to parallel transport in the geometrical world. We preferred to state the second assertion as a hypothesis since the physical parallel transport might be also defined by some other method seemingly different from the method used in our hypothesis.

The first and second hypothesis impose a relatively small number of restrictions on the geometry of the world. The greatest number of restrictions is imposed by third hypothesis:

3) *There exist certain relations between the fundamental metric tensor and the quantities characterizing matter (gravitating masses, electromagnetic phenomena, and so on), which we call world equations.*

In order to define properly the third hypothesis, it is necessary to determine which quantities characterize matter and to write down their relations which we called world equations. However, it is impossible to do this because deriving the world equations, and even their simple explanation, requires rather complicated mathematical apparatus. To describe somehow the general nature of the world equations we will make several remarks. The world equations are differential equations, whose unknown functions are, on the one hand, the quantities g_{ik}, the components of the fundamental metrical tensor, and, on the other hand, those quantities which characterize matter. The world coordinates x_1, x_2, x_3, x_4 are independent variables in these equations.

The postulate of invariance requires that the world equations represent the intrinsic properties of the world and possess, to some extent, invariant nature under transformations of the world coordinates. The last requirement imposes rigid restrictions on the arbitrariness of the world equations and considerably justifies their form, which is adopted in the third hypothesis. Like any equations of the natural sciences, the world equations can be considered in greater or lesser detail. This greater or lesser detail of the consideration of the world equations affects mainly that part of them which depends on matter. If of all matter we consider only the gravitating masses, then the world equations will not contain quantities characterizing the electromagnetic processes in the world. The form of the world equations in this case will be such that, as a consequence of them, the worldline of a free material point will be a straight line; this should be expected since, if this were case, the second hypothesis would contradict the third. In particular, if the gravitating masses are stationary, then the only quantity which, on the part of matter, enters the world equations will be density $\rho = \theta(x_1, x_2, x_3, x_4)$.

If we accept, on the basis of the electromagnetic theory of matter, that all matter reduces to electromagnetic processes, then the quantities characterizing matter in our equations will be: electrical density, three components of the electrical current, electrostatic potential and three components of the electromagnetic (vector) potential – altogether eight quantities. Further, accepting Mie's view on the theory of matter, all electromagnetic processes can be reduced to the electromagnetic field, and then only four quantities characterizing matter will enter the world equations – the electrostatic potential and the three components of the electromagnetic potential. In this case, as Hilbert showed, a consequence of the world equations will be Maxwell's equations which govern all electromagnetic phenomena.

How are the world equations found? Due to technical difficulties, this question cannot be answered here in detail. But it is necessary to say several words about it. The heuristic method of establishing the world equations is to obtain them from the so-called *variational principles*, which are analogous to the principle of the least action in mechanics. It is necessary to be careful not to see in these principles something more than the heuristic method; it is, of course, tempting to link them with teleological ideas, but this at the present time cannot be done, even formally. The essence of this principles is that one chooses a certain expression which is of great importance for the geometry of the world (for example, the mean curvature), and tries to determine its minimum. The world equations are necessary, but insufficient conditions for this minimum. As these equations are insufficient for the chosen expression to be a minimum, then one should not make, the tempting from a teleological point of view, assertion about a structure of the world with the smallest curvature and so on. The heuristic method of deriving the world equations, of course, cannot serve as proof of their correctness. The world equations are formulated like any law of nature. Their formulation is a hypothesis, which should, after a corresponding interpretation, be tested experimentally. Let us see how the above three hypotheses, which we will call *hypotheses of gravitation*, can be probed by experiment and how we can interpret them in order to test them.

4. It is always possible to choose coordinates of the world in such a way that at a given point and in its small vicinity[61] the world is with Euclidean metric and Euclidean properties. These coordinates are called Riemannian coordinates. Riemannian space coordinates of

---

[61]The expression "in its vicinity" should be understood in the sense that the properties of the world, with the chosen coordinates, will be close to Euclidean, not only at the given point, but in its neighborhood as well. It is impossible to give here a rigorous mathematical formulation of this assertion.

the given point can be represented by the usual rectilinear, orthogonal coordinates. We will use these rectilinear, orthogonal coordinates not only for the given point, but also for its vicinity; more definitely, we will assume that the vicinity of a point has Euclidean metric and will regard the introduced space coordinates as rectilinear, orthogonal coordinates, whereas sidereal time will be taken to be the time coordinate. It is self-evident that the assumption of the Euclidean metric will be incorrect, but the smaller the vicinity of a point is, the closer the geometry of the world around this point to the Euclidean world is. We will call the world constructed in such a way, *conditional* world. A material point in this conditional world, which is subject to the influence of gravitating masses, will have a worldline different from the straight line in the conditional world. In fact, according to the second hypothesis of gravitation this worldline will be a straight line in our physical world, which is different from the just introduced conditional world, and therefore that worldline will differ from the straight line in the conditional world. In this way, a point in the conditional world will have acceleration, and therefore, by the principle of inertia of the old mechanics we can determine the force acting on the point, which is induced by the gravitating masses and is acting in accordance with the laws of the old mechanics. In other words, we can determine the force of universal gravitation of the old mechanics. This force will depend, of course, on the gravitating masses because it is determined from the fact that the worldline of the point will be a straight line in the physical world and, therefore, a certain curve in the conditional world, whose shape will depend on the metric of the physical world, i.e., by the third hypothesis, on matter and, in particular, on the gravitating masses. What should we expect to have in view of these considerations? As a first approximation, we should get Newton's law, after that we should, generally speaking, get a correction to the universal law of gravitation as greater as closer to huge gravitating masses the material point is.

By studying the motion of a point around the gravitating mass of the Sun, Einstein derived as a first approximation, as expected, Newton's law; but in addition to this law there was a certain correction. The use of this correction gave just that residual motion of the perihelion of Mercury, which was unexplained. The unexplained part of the motion of the perihelion was $43''$ in 100 years; Einstein's calculations gave the value $42'', 89$ in 100 years; closeness of these two numbers is especially remarkable when we remember how we came in a roundabout way to the second number.[62]

---

[62]I presented here the usual point of view allowing, through the use of Euclidean geometry and the old mechanics, to find experimental relations in the physical

The path for further verification of the hypothesis of gravitation was opened. Changing the metric of the world, a gravitating mass should influence almost all physical phenomena. For example, light passing near a gravitating mass and having as worldlines special straight lines of the physical world (the so-called null lines), in the conditional world it should deviate from the rectilinear propagation in the conditional space and should be curved as a result of the action of the gravitating mass. The expedition, equipped for the observation of the last a total solar eclipse in May 1919, found a shift in the star locations through the star light which passes near the Sun, whose image is covered by the Moon. The observed shift was in perfect agreement with that calculated by Einstein.

The periods of oscillations of electrons on the Sun should be longer than those on Earth as a result of the change of the metric of the world induced by the Sun's huge gravitating mass (since the metric affects time intervals as well). Calculations show that this lengthening will affect the shift of the dark lines of the spectrum toward its red end by 2×10^{-6} wavelengths of the line. The observation gives an average number for the shift equal to 1.85×10^{-6} wavelengths. There exists a number of experimental facts which confirm the correctness of the discussed hypothesis of gravitation and so the correctness of Einstein's theory known as the *general principle of relativity*. It should be nev-

---

space, which serve as verification of the hypothesis of gravitation. Unfortunately, this usual point of view is devoid of sufficient clarity and rigor. It would be more correct to reason in the following way. A geometrical quantity in a region of the world, around a given point (event), is expanded in a series of powers of the differences of the world coordinates of the points in the area and the world coordinates of the given point. Considering only the most rough approximation, we obtain the Euclidean geometry of space, the old mechanics and Newton's law. Going further in our approximations, we should obtain a correction to the Euclidean geometry and the old mechanics as well as to the law of gravitation. Using the conditional world, we regard the correction to the Euclidean geometry as a higher order of smallness (smaller) than the correction to the law of gravitation, thus we obtain the motion of the perihelion of Mercury. Here we should point out the danger encountered often in approximate calculations and considerations, namely the danger of an explanation of certain phenomena by keeping some small quantities and neglecting others of the same order of magnitude that could destroy the influence of the first. From this perspective, the question remains unanswered in the theory of relativity. Moreover, no attention is paid to it, probably here no difficulties are encountered. But if we imagine for a moment that when we are setting our approximate considerations we act wrongly and neglect terms of some order of magnitude while retaining terms of the same order of magnitude, it may happen that no motion of the perihelion of Mercury can be explained. Let me repeat that this is not very likely, but I still found it useful to draw your attention to this consideration, which needs more detailed clarification and which contains some danger for the possibility of experimental verification of the theory of relativity.

ertheless pointed out that this correctness has been so far confirmed only very roughly. The principle of relativity should not be regarded as something completely confirmed. Of course, many details of Einstein's theory, particularly those dealing with the formulation of world laws, can and should change as a result of new experimental facts and the constant advancement of mathematical analysis. What is important is the fact that the general features of Einstein's theory brilliantly withstood the experimental tests, not only by explaining what seemed inexplicable, but also by predicting, following the example of the classical theories, a number of new phenomena.

§10 Matter and Structure of the Universe

1. In Einstein's theory, electromagnetic phenomena are expressed by special quantities, which do not represent any property of the geometrical world. The German mathematician Weyl, developing the theory of Einstein, created a system, where the electromagnetic phenomena of the physical world are represented, with the help of some additional assumptions (taking into account Mie's theory), by certain properties of the geometrical world. These additional assumptions are not so essential and a generalized geometry could get rid of them[63]. An essential feature of Weyl's theory is the grandiosity of his conception, which interprets everything in the physical world as some properties of the geometrical world. Let us see what kind of hypotheses characterized Weyl's theory.

Its first theoretical advantage over the hypothesis of gravitation is reducing the number of hypotheses giving them a more general character. The restrictions requiring the geometry of the world to be Riemannian geometry are not present in Weyl's theory. At the same time the motion by inertia is includes a wider range of motions. The third hypothesis of gravitation changes in the sense that in its formulation, according to the general idea of the theory, there is no indication whatsoever of the role of the quantity, which represents matter. The hypotheses of Weyl, which, for brevity, we will call *hypotheses of matter*, are formulated in the following way:

1) *A material point subject to the action of a gravitating mass as well as to the action of electromagnetic phenomena moves by inertia.*

2) *There are given relations between the fundamental metric tensor and the scale vector, which called world equations.*

---

[63]Eddington, Schouten and the author of the present work have papers on this topic.

If there is nothing in the physical world except electromagnetic phenomena and gravitating masses, or if everything reduces only to electromagnetic phenomena, then the first hypothesis can be expressed simply by requiring that every motion would be motion by inertia (in the sense of the new mechanics, of course). It follows from the combination of the second and the first hypothesis that all quantities representing electromagnetic phenomena (assuming here that Mie's theory of matter is correct) dependent on the geometrical quantities of the geometrical world.

In fact, the life of every material point is defined by a straight worldline, and a straight line, in turn, depends only on the fundamental metric tensor and the scale vector, which, by the second hypothesis, are connected only with each other and nothing else. Weyl makes a more definite assumption by pointing out that the four quantities defining the scalar vector represent electromagnetic and electrostatic potentials in the physical world. This specific assumption, however, is not essentially necessary for the Weyl's theory.

The world equations are derived in a way that is analogous to the derivation of the world equations in the theory of Einstein. The only essential distinction of the world equations in the Weyl's theory is the more rigid requirement of their invariance. The additional requirement of *scale invariance* is added to the ordinary principle of invariance. We will consider these additional requirements; we will call *change of scale* (see above §3) the operation of changing the magnitude of infinitely small vectors, which is different for different points; in other words, the operation of multiplying the components of the fundamental metric tensor g_{jk} by the quantity μ, which is a function of the coordinates x_1, x_2, x_3, x_4. There is nothing new or extraordinary in this operation: for example, in different countries different scales are used, i.e. in Russia – archine, in Germany – meter, in England – foot. Let us imagine that we have to change the scale from point to point, then we will have the operation of *changing the scale*, described above. Changing the scale in the geometrical world will correspond to different ways of measuring length (magnitudes of vectors) in the physical world. These different ways will use the same initial value, but then will change the unit of measurement from point to point. In this way, the change of scale in the physical world will correspond to a change, from point to point, of the unit of measuring length, which change, as we explained in §3 always takes place.

Intrinsic properties of the world are divided in two classes. Those belonging to the first class do not depend on the above mentioned changing of scale, better said, do not change their form when the scale changes. The intrinsic properties of the second class change their form when the scale changes. We call the intrinsic properties of the world, belonging to the first class, *scale-invariant*.

Weyl extends the postulate of invariance by adding to it the requirement that *all physical laws should be not only intrinsic, but also scale-invariant properties of physical world.*

In agreement with such an extension of the postulate of invariance, it is necessary to require that the world equations be expressed in a form which is not only coordinate-invariant but also scale-invariant.

2. An important difference of the theory of Weyl from the theory of Einstein is the absence of experimental proof of Weyls' theory. I think that the requirement of accepting the theory of Mie is not essential for the general ideas of Weyl's theory; but if we take Weyl's theory as it is presented by him, then the acceptance of Mie's theory is unavoidable; but that theory is not in agreement with the experimental facts. In any case, Weyl's hypothesis of matter still awaits its experimental confirmation.

Some essential support of Weyl's theory comes from the fact that Maxwell's equations are derived from Weyl's world equations taken even in their most general form. The law of conservation of energy follows from both theories – Einstein's and Weyl's, which provides additional support for Weyl's theory. The future will decide whether Weyl's theory it its present form will be experimentally confirmed or it will need some new changes and additions. But this is not so important. What is important in Weyl's theory is its assertion that *all material phenomena of the physical world favour only certain interpretations of properties and images of the geometrical world.*

Let us consider some consequences of the hypothesis of gravitation and the hypothesis of matter, which will give us better understanding of the interpretation of matter by the geometrical properties of the world. The gravitating mass is connected with the vectorial curvature; the greater the gravitating mass, the greater the vectorial curvature. The electromagnetic phenomena are connected, on the contrary, with the metrical curvature; the more intensive the electromagnetic phenomena, the greater the metrical curvature. The vectorial curvature around the Sun[64] is significantly greater than in the interplanetary

---

[64]This expression is not completely right; we should not forget that both the vectorial and the metrical curvatures belong to the *four-dimensional* world. However, we will intentionally use imprecise expressions in order to achieve greater visualization.

In essence, we have to consider the curvature of the hypersurface of the four-dimensional world $x_4 = x_{40}$, i.e., the curvature of the space corresponding to the moment x_{40}

space, far away from any stars; the metrical curvature around a powerful generator is significantly greater than the metrical curvature far away from large masses of electromagnetic energy; the metrical curvature in a ray of powerful projector is much stronger than the curvature in a ray of a lamp. The considerable vectorial curvature around the Sun can be detected when a vector is parallel transported along a closed curve and we monitor its direction. The metrical curvature, around a powerful generator, can be detected through a significant change of the length of a vector which is parallel transported along a closed curve around the generator. Thus, our physical world is characterized in its different parts by a great diversity of vectorial and metric curvature. Our space is a special hypersurface of the world; the space has both vectorial and metrical curvature, which are extremely different in its different parts and at the same time change together in time. In order that we monitor all the time the changing geometrical properties of the material space, we need a very detailed information about the evolution of the material space. Although we do not have this information it is extremely interesting to try to discover the properties of space in which we live and in which the stellar bodies move. Further we will talk a bit about the structure of the Universe.[65]

3. Approaching the issue of the structure of the Universe, first of all, we have to keep in mind that our Universe, i.e., the material space, where stars move, does not exist independently on its own, and can be regarded only as a hypersurface of the world which corresponds to a given values of the time coordinate. Speaking of the geometrical properties of the Universe, we have to, based on the discussion above, determine first the geometrical properties of the world, and next consider the hypersurfaces in this world, which correspond to different values of the time coordinate, and then study the geometry of these hypersurfaces.

The geometrical properties of the world, whose interpretation is the physical world, according to both the hypothesis of gravitation and the hypothesis of matter, are fully determined as soon as we have information about the matter filling this world, in other words, as soon as we know the matter filling the physical space and its motion in time. The difficulty of solving the issue of the geometric properties of the world in general terms forces us to make a number of simplifying assumptions related primarily to the properties of matter that fills the

---

[65]EDITOR'S NOTE: The expectations expressed in this section have not been confirmed because the efforts to reduce electromagnetic phenomena to geometrical properties of the space were unsuccessful. See W. Pauli "Theory of relativity" for more about the theories of Mie and Weyl.

world.

The first of these simplifying assumptions, accepted in the present time, in the study of the Universe, based on the theory of Einstein, completely ignores the electromagnetic phenomena and reduces all matter, filling the world, to gravitating masses. Then taking into account the hypothesis of gravitation, the world equations make it possible, by using quantities, characterizing the gravitating masses in the world, to determine g_{jk}, i.e., the fundamental metrical tensor of the world and therefore the metric and all geometrical properties of the world connected with it. The quantities characterizing the matter in the world, when it is reduced to gravitating masses, are the unknown density of the gravitating masses $\rho = \theta(x_1, x_2, x_3, x_4)$ and those data which determine the worldlines of these gravitating masses. As indicated above, as a consequence of the world equations we will obtain the worldlines of the gravitating masses among the straight lines of the world. Without going into detail I will remark only that the data determining the worldlines of the gravitating masses will boil down at the end to three unknown functions of the coordinates of the world. In this way, ten world equations should be used to determine fourteen unknown quantities: ten components of the fundamental metric tensor, one density and three quantities determining the 'life' of the gravitating masses. As we could be interested only in the intrinsic properties of the world we can replace one coordinate system of the world with another. Then, as it is proved in deferential geometry, out of the ten components of the fundamental metric tensor we can in advance assign, depending on the coordinates, certain values to four components with the help of suitable coordinate transformations. Thus, of all components of the fundamental metric tensor only six components $(10 - 4)$ should be determined; so the ten world equations will be used to determine ten $(6 + 1 + 3)$ unknown functions of the coordinates and the time. Then the solution of the problem of determining the geometry of the world becomes possible, at least in principle. If we determine the ten functions, then we will know both the geometry of the world and the distribution of the gravitating masses together with their worldlines. It should be pointed out that up to now nobody has solved the above problem in its entirety and the possibility to determine the ten functions from the ten world equations has not been studied yet. It is even possible to argue, on the basis of the existing studies of the world equations, that it is impossible to obtain unambiguously the ten quantities, mentioned above, from the world equations. Thus, we will need either some additional principles for defining these quantities or using astronomical data in order to define the task of studying the

Universe more clearly.

If, however, we manage to find any form, even some special form, of these ten functions, then with more detailed astronomical data about our Universe, we could have excellent and detailed method of experimental verification of Einstein's theory and of the discussed solution of our task. In fact, our world equations would determine the density of the gravitating masses, in other words, we would have *the distribution of the gravitating mass in the physical space* at any moment of time. And, finally, having determined the worldlines of the gravitating masses, our equations would give us information about the motion of the gravitating masses in space at any moment of time. In other words, our equations would give us at any moment of time *the distribution of the velocities of of the gravitating masses in the physical space*. The distribution of the gravitating masses could be determined precisely with the help of astronomical observations of the distribution of the stars in the Universe. The distribution of the velocities would be tested in the same way – by astronomical observations of the velocities of the stars in the observable Universe.

Having tried all special solutions of the world equations, we, with the help of the astronomical data, would determine the geometry of the world and the distribution and the 'life' of the gravitating masses of the Universe.

Unfortunately, the outlined, ideally correct approach, is not practically applicable. On the one hand, the mathematical study of the ten world equations could not yet be carried out completely. On the other hand, the astronomical data are not sufficient for the discussed experimental verification.

4. It is necessary to find another approach. For simplifying the mathematical aspect of the task we have to make arbitrary hypotheses, which reduce the world equations to simpler ones. For the experimental verification we should deduce some simple consequences from the structure of the Universe, suggested by the study of simplified world equations, and compare these simple consequences with the available astronomical data and verify in this way the experimentally developed mathematical theory. Based on what we said, we have to consider the study of the Universe as a whole at the present time in its very infant stage of development. All conclusions, without exception, deduced from the study of the Universe at the present time, should be treated with complete distrust; This distrust is also supported by the extreme unsteadiness and unreliability of our astronomical data about the Universe.

Simplifying assumptions, used at the present time as a foundation for the study of the Universe, concern two sides of the issue: first of all, simplifying assumptions regarding gravitating masses are made; *the relative velocities of the gravitating masses are considered equal to zero*, the gravitating masses are considered stationary. At first sight, these assumptions seem absurd, but we have to remember – as the velocities of the gravitating masses, as the observations show, in most cases are extremely small compared to the speed of light, we can with high accuracy represent the world equations in such a form where the velocities are indeed equal to zero. Concerning the density of the gravitating masses no additional assumptions should be made.

The second class of assumptions deals with the geometry of the world. It is assumed, fist of all, that the geometry of the world has the property to define a space (hypersurface), in which the curvature is the same in each of its points and changes only with time. Further, one more additional assumption about the metric of the world is made – this assumption simplifies the calculations considerably, but its meaning is not clear so far; as I do not like to go into details and I will not discuss this assumption.

Having made the above assumptions it is possible to arrive at two types of Universe:[66] 1) *stationary type* – the curvature of space does not change with time, and 2)*non-stationary type* – the curvature of space changes with time. A sphere, whose radius does not change with time, can serve as an illustration of the first type (stationary) Universe. The two-dimensional surface of the sphere is a two-dimensional space of constant curvature. Conversely, the second type of Universe can be pictured as a sphere which changes all the time – either expanding or reducing its radius and as if contracting. The stationary type of Universe allows only two cases for the Universe, which were studied by Einstein and de Sitter. Einstein determined, using the existing astronomical data, that the radius of the curvature of the Universe is $10^{12} - 10^{13}$ distances from the Earth to the Sun, and that the density φ (everywhere constant) is 10^{-26} g/cm^3. In de Sitter's model of the Universe there is complete absence of density of gravitating masses ($\rho = 0$).

The non-stationary type of Universe presents a great variety of cases: for this type there may exist cases when the radius of the curvature of the world, starting from some magnitude, constantly increases with time; there may further exist cases when the radius of curvature changes periodically: the Universe contracts into a point (into nothingness), then again, increases its radius from a point to a given

---

[66]The question of the structure of the Universe is considered by Einstein, De Sitter and others.

magnitude, further again reduces the radius of its curvature, turns into a point and so on. This unwittingly brings to mind the saga of the Hindu mythology about the periods of life; there also appears a possibility to speak about "the creation of the world from nothing," but all of that should be viewed as curious facts which cannot be solidly confirmed by the insufficient astronomical material. In the absence of reliable astronomical data, it is useless to give any numbers characterizing the "life" of the non-stationary Universe. If nevertheless, for the sake of curiosity, we try to calculate the time elapsed from the moment when the Universe was created starting from a point to its present state, that is, when we try to determine therefore the time that elapsed from the creation of the world, then we obtain a number in the tens of billions of our ordinary years.[67]

In the conclusion of the discussion of the structure of the Universe, we will mention a misunderstanding, which is repeated not only in popular articles and books, but also in more serious and specialized works on the principle of relativity. I mean the notorious question of *the finiteness of the Universe*, i.e., the finiteness of our physical shining with stars space. Some claim that a constant and continuous curvature of the Universe leads as if to the conclusion that it is finite and, first of all, that a straight line in the Universe has "a finite length", that the

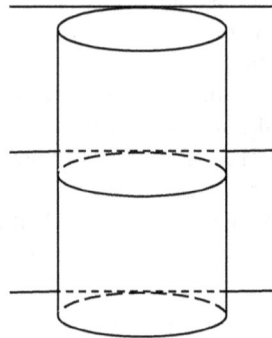

Figure 22

volume of the Universe is also finite, etc. This claim may be based either on misunderstanding or on additional hypotheses. *It does not follow from the metric of the world at all*, and only the metric can be elucidated by the world equations. Simple examples can convince us in it. The metric of the surface of the cylinder and the metric of the plane are the same, and on the cylinder do exist "straight" lines of finite length (a circle, see Fig. 22), whereas such straight lines do not exist at all on the plane. The issue of the finiteness of space depends not only on its metric but also on the requirement of how two coordinate systems define the same point. To a great extent this requirement is arbitrary; even its rational restriction, about which we cannot talk

---

[67]EDITOR'S NOTE: It should be specifically pointed out that Friedmann explored the hypothesis of a non-stationary world only on the basis of the logical structure of general relativity. There existed no experimental fact that might be interpreted as a manifestation of such a non-stationary Universe; the recession of nebula was discovered by Hubble in 1924-1927.

here, leaves sufficient room for the arbitrariness of the sameness and distinguishability of the points of space.

Thus, the metric of the world alone does not give us any possibility to resolve the question of the *finiteness* of the Universe. For the solution of this question we will need additional theoretical and experimental studies. Of course, with suitable conventions the question of studying the finiteness of the Universe is not hopeless. Suppose that shadows living on the surface of a sphere decided to investigate this question. They could solve this problem by sending someone to travel on the sphere. Keeping the direction of a straight line all the time, and following it in the same direction, our traveller, observing the features of the surroundings, would see that they are constantly changing during the time of his journey. He could come across other landscapes and spherical towns very little resembling his native country. But approaching, from the other end, his town from where he started his journey, the traveller would notice that the surroundings start to resemble more and more those which he left. Having come back to the initial point from which he started his journey, the traveller, trough careful observations, could come to the conclusion that the point where he just arrived completely coincides with the point from where he left. Thus, the issue of the finiteness of the Universe of the sphere could be, of course, resolved by a number of conventions and additional studies, but the metric of the sphere alone could not resolve the issue of its finiteness.

So, it does not follow from the constancy and positiveness of the curvature of the Universe at all that our Universe is finite.

§11 General Conclusions of the Principle of Relativity

1. It seems worthwhile to summarize the conclusions to which we were led by the relativity principle. I will do it in this section by mentioning only the most important facts, leaving out all details and the corresponding mathematical apparatus.

A space of any dimensions can be arbitrarily arithmetized in such a way that every point of the space is assigned a combination of several numbers, called *coordinates* of the given point of space. The method of arithmetization is completely arbitrary.

The properties of space can be *intrinsic*, i.e., not depending on the arithmetization of space, or *extrinsic*. The intrinsic properties are invariant under a transition from one method of arithmetizing space to another. The intrinsic properties of space do not depend on our will, i.e., on the choice of a method to arithmetize space. The extrinsic properties are rather properties of a given arithmetization of space.

In space we define *a distance* between two points and, respectively, a magnitute of an *infinitesimally small vector at a given point of space*. The latter quantity depends on the *fundamental metric tensor*. In the four-dimensional space the distance is called *interval* and the fundamental metric tensor depends on *ten quantities* g_{ik}.

In space we also define a *parallel transport of a vector* which serves for defining *straight lines* in space and its *curvature*. The parallel transport of a vector depends not only on the fundamental metric tensor, but also on certain quantities, called *scale vectors*; in a four-dimensional space a scalar vector is defined by four quantities. The change of the direction of a vector, when it is parallel transported, forms the basis of the notion *vectorial curvature of space*, and the change of the magnitude of a vector defines *metric curvature of space*.

If the magnitude of a vector does not change when it is parallel transported then the space is called *Riemannian*. In the Riemannian space the scale vector is absent, the metric curvature is equal to zero, and the straight line is the shortest. Other spaces are called Weyl's spaces and in them the straight line is not the shortest.

2.   In addition to the geometrical space we introduce a *three-dimensional physical space* and *a physical time* – and we introduce them not separately, when their physical existence is unthinkable, but in their inseparability which constitutes the physical world. The physical world consists *of matter* (interpreting this word in the broadest sense); both gravitating masses and electromagnetic processes belong to matter. I would like to stress that namely the world consists of matter, because matter in space and without time is physically unthinkable. We will adopt a special interpretation of the geometrical world (the space of four dimensions) with the help of the physical world. Every object of the geometrical world is interpreted by a an object in the (material) physical world. *This interpretation is completely arbitrary and depends on our will*. We will use an interpretation in which the physical world serves as an interpretation of the four-dimensional space; we will call this four-dimensional space simply *geometrical world*. In the physical world we identify a group of properties, which we call *physical laws*. For these physical laws we establish *the postulate of invariance* consisting in the requirement that the physical laws are invariant under the transition from one method of aritmethization to another, and represent therefore *the intrinsic properties* of the geometrical world.

From a certain point of view, the postulate of invariance could serve as a definition of physical laws.

We put three coordinates of the physical world in a special group called *space coordinates*, and the fourth coordinate is called a *time coordinate*. For the metric of the geometrical world we set a special *postulate of realness*, whose essence is related to certain properties of realness and imaginariness, which should characterize the interval between two simultaneous points and two points having the same space coordinates. The postulate of realness imposes restricting conditions not only on the metric of the world, but on the arbitrariness of its arithmetization, because in each arithmatizion, due to the postulate of realness, it should be possible to distinguish the time coordinate. The postulate of realness plays very important role in the so called principle of causality. According to this postulate, two events, happening at different times, could not be made simultaneous with the help of arithmetization obeying the postulate realness. In this way, cause and effect in one method of aritmetization of the world also remain cause and reason in another method of introducing world coordinates.

The change of space coordinates with time we call *motion*. Motion is characterized and determined by *the worldline of a point*.

When a conditional interpretation of the geometrical world, with the help of the physical world, is adopted, then it becomes possible *to determine the geometry of the geometrical world through experimental studies of the physical world*. Simple by idea, but difficult by realization is the experimental determination of the intervals in the physical world; it gives us the metric of the geometrical world. The experimental study of the change of the length of a vector when it is parallel transported, makes it possible to determine the scalar vector of the geometrical world. But except the mentioned simple by idea experiments, it is possible, with the help of experimental studies of any intrinsic properties of the physical world, to make a certain conclusion about the geometrical world. Each physical law, according to the postulate of invariance, defines a certain intrinsic property of the geometrical world. Studying experimentally any physical law we thereby establish a given property of the geometrical world. Newton's law of universal gravitation was used by Einstein to explain the metric of the world, whereas the laws of electrodynamics allowed Weyl to determine not only the metric of the world but also its scale vector.

3. The principle of inertia divides the motion of material points into two classes. The motion of the first class is represented by straight worldlines – the straight lines in the geometrical world; this is motion *by inertia*. The motion of the other class is not represented by straight worldlines – this is motion *under the action of forces*. A force involved

in such a motion is characterized by a special number – the *mass* of a material point, distinguishing one material point from another and *quantities* (compare accelerations) which represent the degree of deviation of the worldline of a particle from the straight line of the world.

The impossibility to study the geometrical world from a general point of view, forces us to make a number of hypotheses about its metric and its properties. The narrowest of these hypothesis is the hypothesis of the old mechanics, according to which the geometrical world *is an Euclidean four-dimensional space.* There is no room for experimental studies of the metric of the world in the old mechanics, because it is defined in advance by the above hypothesis. Matter, consisting of gravitating masses, from the point of view of the old mechanics, possesses the property of universal gravitation. Matter, consisting of electromagnetic processes, is governed by the equations of electrodynamics. Unlike the hypothesis of the geometry of the world, adopted by the old mechanics, the theory of Einstein employs a much broader hypothesis which we called *the hypothesis of gravitation.* The geometrical world presupposes not Euclidean but a wider geometry – Riemannian. *The existence of gravitating masses causes only motion by inertia*, and therefore the force of the universal gravitation is only an apparent force. Finally, the metric of space is linked to quantities, characterizing matter, through special *world equations.* By studying the motion of gravitating masses through experiment we can determine the metric of the geometrical world.

The theory of Weyl goes even further than the theory of Einstein. The hypothesis of Weyl, which we called *the hypothesis of matter*, is the following. According to the hypothesis of matter (in the presence of gravitating masses and electromagnetic processes), every motion is motion by inertia, and the metric of the world and its scale vector are defined by special world equations. All properties of matter (consisting of electromagnetic processes) are obtained from the geometrical properties of the world. *There is nothing except these geometrical properties of the world.*

The theory of Einstein is proved by experiment. It explains the old, seemingly unexplainable, phenomena and predicts new striking effects. The most truthful and deepest method of studying, with the help Einstein's theory, the geometry of the world and the structure of our Universe is the application of this theory to the entire world and to the employment of astronomical studies. So far this method gives us a little, because the mathematical analysis puts down its weapons before the difficulties of this issue, and the astronomical studies do yet not give a sufficiently reliable basis for the experimental study of our Universe. But in these circumstances, we should not see only temporary difficulties; our descendants will no doubt discover the nature of the Universe in which we are destined to live ... Yet it seems that[68]

> It seems to us still, that
> To measure deep ocean,
> Calculate sands, rays of planets,
> Even if high mind could do it,
> There is here no number or limit!

---

[68]EDITOR'S NOTE: Translated by Sergey Andronenko, St. Petersburg, Russia.

www.ingramcontent.com/pod-product-compliance
Lightning Source LLC
Chambersburg PA
CBHW071105210326
41519CB00020B/6171